ADVANCES IN LIBRARY ADMINISTRATION AND ORGANIZATION

ADVANCES IN LIBRARY ADMINISTRATION AND ORGANIZATION

Series Editors: Edward D. Garten and
Delmus E. Williams

Recent Volumes:

Volume 1:   Edited by W. Carl Jackson,
                   Bernard Kreissman and Gerard B. McCabe

Volumes 2–12:   Edited by Bernard Kreissman and
                   Gerard B. McCabe

Volumes 13–17:   Edited by Edward D. Garten and
                   Delmus E. Williams

ADVANCES IN LIBRARY ADMINISTRATION
AND ORGANIZATION   VOLUME 18

ADVANCES IN LIBRARY ADMINISTRATION AND ORGANIZATION

EDITED BY

EDWARD D. GARTEN

Dean of Libraries and Information Technologies,
University of Dayton, OH, USA

DELMUS E. WILLIAMS

Dean of University Libraries,
University of Akron, OH, USA

2001

JAI
An Imprint of Elsevier Science

Amsterdam – London – New York – Oxford – Paris – Shannon – Tokyo

ELSEVIER SCIENCE Ltd
The Boulevard, Langford Lane
Kidlington, Oxford OX5 1GB, UK

First edition 2001

Library of Congress Cataloging in Publication Data
A catalog record from the Library of Congress has been applied for.

British Library Cataloguing in Publication Data
A catalogue record from the British Library has been applied for.

ISBN: 0-7623-0718-8
ISSN: 0732-0671 (Series)

∞The paper used in this publication meets the requirements of ANSI/NISO Z39.48-1992 (Permanence of Paper).
Printed in The Netherlands.

CONTENTS

INTRODUCTION

Shortly after my co-editor and I assumed responsibility for this series, we had agreed to spend some time talking about our hopes and expectations for the series. We'd agreed to meet in San Francisco prior to a conference, rent a car, drive down Highway 1 as far as San Simeon, and talk. The beauty of that drive, however, quickly pushed any talk of practical things like library administration far from our thoughts. On our return to San Francisco, barreling through the dark, late at night on Highway 101 north, I noticed that we were almost out of gas; indeed the low fuel light probably had come on miles earlier. We were in the midst of nowhere – no sign of a gas station, a town, or even lights in the distance. Then the great Eagles' ballad, *Desperado*, came on the radio. I've been accused of "runnin' on empty" several times in my life, but I didn't relish the thought of spending the night on the side of the road. Off to the left we noticed what likely were the lights of a small village (lights are always further away than they seem). As we made our exit, the Eagles' *Take It To the Limit* came on the radio – "put me on a highway, show me a sign, and take it to the limit one more time". As we entered town several miles later we saw a solitary gas station at the end of Main Street, it's lights just being turned out as we rolled up to a pump (likely on fumes alone). The attendant turned on the pumps, filled our car with gas, and we were (whew!), again, on our way back to San Francisco.

Given the rapidity of change, more than a few times during the last decade more than a few library administrators have been charged with "runnin' on empty"; but more often than not, these same individuals were leading the charge, "takin' it to the limit", stretching the boundaries of practice informed by theory and reflection. In this volume we continue this series' long practice of bringing to its professional and academic readership an eclectic mix of scholarship and longish essays. Included in this volume are four papers given as part of a symposium to honor the career of Richard Dougherty, one of academic librarianships stellar lights. As Bill Gosling notes in this introduction to these four essays, the papers together capture a sense of the forward thinking vision that seeks to reach beyond current practice; that seeks to further innovation. In the two decades that closed the 20th century, library administrators encountered tremendous demands for change and reacted to those demands in responsible, practical, but also, visionary fashion. They truly took it to the limits as they faced the year 2000.

Even with innovative practice there are some issues that continue; with some accentuated. Workplace stress is one of those issues we face increasingly. In the first chapter of this volume Connie Van Fleet and Danny P. Wallace discuss the stress on staff as a result of the very real changes that are occurring at the reference desk as we develop more virtual libraries. In Volume 17 we presented a first-ever for this series contribution on the emerging chief information officer role. In this volume Stephan R. Reynolds as a CIO offers a compelling approach for developing an integrated information resources strategic plan. Automated systems, especially in our academic libraries, have prompted library administrators to evaluate and reconsider their hiring practices and the manner in which they allocate staff time and in his contribution, Murle Kenerson looks at these issues in the context of Tennessee academic libraries. Sandy Slade explores some international perspectives in quality assurance for library support of distance education while Theresa Byrd's contribution to perspectives on quality assurance offers practical insight into the implementation of a Total Quality Management system in a group of community college libraries.

As we go forward into the next period of librarianship, some things remain unresolved and vexing. One of these unresolved issues relates to compensation for academic librarians. While gender in academic librarianship remains predominately female, gender-based wage disparities continue to exist among men and women in the profession. Elizabeth A. Titus offers readers a way to determine and measure gender-based salary disparities of individuals in academic libraries. Finally, each of the contributors to the Dougherty Symposium, held in 1999 at the University of Michigan, are well-regarded library administrators or academic administrators. Many of the observations presented in these essays are provocative and challenging. Clearly, our work in academic libraries is sited within the complex and often puzzling culture of academe. These papers note some of the obstacles as well as some of the challenges that, together, we face in our attempts to clear the way for fresh visions of library leadership suited for the years ahead.

Edward D. Garten
Co-editor

VIRTUAL LIBRARIES – REAL STRESS: CHANGE AT THE REFERENCE DESK

Connie Van Fleet and Danny P. Wallace

TECHNOSTRESS DEFINED

The demands placed on libraries, library employees, and library patrons by the seemingly rapid introduction of computers into the library equation has created an intensity of activity and demand that appears to greatly exceed that of previous generations of library practice. Although providing patron services in the library context has always been stress-prone, the influence of digital media has resulted in an epidemic of what Brod (1984, 16) termed *technostress*: "a modern disease of adaptation caused by an inability to cope with the new computer technologies in a healthy manner".

CHANGE AND SOCIETY

Technostress has many root causes, one of which is change itself. The very real concern over the realities of change is reflected in the substantial popular literature of change. Bell (1973) referred to the impact of "post-industrial society". Toffler, one of the most influential and prolific of popular commentators on change (1970, 1980, 1990) wrote of the psychological and sociological disruption caused by "future shock", the social revolution of the "third wave", and the disruption to be anticipated with a major societal "powershift".

Knowles (1980) described the stress-intensive impact of change on individual human beings in terms of the relationship between the time-span of social

Advances in Library Administration and Organization, Volume 18, pages 1–44.
2001 by Elsevier Science Ltd.
ISBN: 0-7623-0718-8

change and human life expectancy (Fig. 1). At the time of ancient Rome, most individuals experienced their entire lives in the context of extremely limited societal change; it was typical for a child to pursue exactly the same course in life as his or her parents. Not until the nineteenth century did the time-span of social change begin to become shorter than the expected individual life-span. By the mid-twentieth century, social change was taking place at so rapid a pace that any individual could expect to experience several major societal transformations during his or her lifetime. Most individuals in current society are employed in multiple successive careers. Even if they remain in the same basic occupation, that occupation changes dramatically enough that it is no longer the career they entered.

Schon (1971) referred to "the loss of the stable state", a situation in which the comforts of perceived personal and societal stability are withdrawn, seemingly abruptly, creating an adaptive challenge for which many individuals are poorly prepared. In this environment there is a pervasive sense that "the anchors of personal identity are everywhere being eroded" (p. 22). Both Knowles and Schon emphasized the role of learning and education in coping with change. According to Schon, "The task which the loss of the stable state makes imperative for the person, for our institutions, for our society as a whole, is to learn about learning" (Schon, 1971, 30). Knowles emphasized the need to explore the ways in which adults learn and can be taught and proposed a system of andragogy to describe adult learning needs. Andragogy is distinctly different from the pre-existing system of pedagogy, which emphasizes the learning needs and tasks of children and techniques for teaching children. Knowles and other humanistic writers "emphasize the importance of experiential learning, of

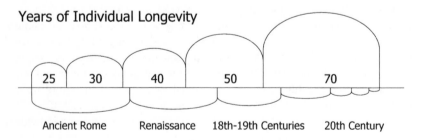

Fig. 1. Time-Span of Social Change vs. Individual Life-Span.
(Adapted from Knowles)

relevance to the individual learner, of respect for the individuality of the learner, of the move toward self-motivation and direction, and of the relationship of the facilitator and the learner" (Van Fleet, 1990, 192).

Lauer (1991) has emphasized that change is very poorly understood even among social scientists who study change. There are many competing theories regarding change as a sociological force, but they are all in agreement that change is an enduring source of concern for both societies and individuals. Although change is not necessarily traumatic and may be generally beneficial, change is resisted by many people for many reasons. Change may be resisted when it is viewed as in some way threatening, when analysis suggests that the costs of change outweigh the benefits, or when the prospective change is perceived as infringing individual or group freedom (Lauer, 1991). Change may also be resisted as an approach to maintaining the integrity of social systems or when the agent of change rather the change itself is perceived negatively (Klein, 1985). When the pace of change is accelerated, resistance to change may become a reflexive response based on the sheer lack of time to assess and assimilate the change.

A BRIEF HISTORY OF TECHNOSOCIETY

Human life since the inauguration of the industrial revolution has been characterized by increasing dependency on technological solutions to societal and individual problems. During much of this era of technological dependence the impact on individuals was sharply focused and very limited. Those people who directly experienced major transitions such as the introduction of the automobile were undoubtedly deeply affected by the transportation revolution. Succeeding generations of drivers, however, have mostly seen rather subtle advances in automotive technology that have had little impact on the basic nature of what it means to drive a car. Factory workers may experience changes in equipment and processes as major disruptions, but those disruptions are generally temporary and generate limited awareness on the part of the general public. It is no wonder that librarians, who have been charged with the preservation and dissemination of knowledge and information, would experience a profound impact in the workplace as modern technology is introduced into the information society.

In more recent years, the direct impact of changing technology – primarily digital electronic technology – has been substantially more pervasive than previously. Although it is difficult to precisely identify the point in time when this major change in the influence of technology occurred, it can to a considerable extent be tied to the introduction of the transistor in 1948. The miniaturization made possible by the transistor revolutionized consumer electronics by reducing both the size and price of electronic products and by improving reliability. The introduction of the microprocessor in 1971 led to further reductions in size and complexity and at the same time allowed for sophisticated computer control of pre-existing products as well as the introduction of completely new digital products.

The ultimate extension of this invasion of digital technology is the personal computer. Individuals born in 1948, the year that saw the introduction of the transistor and Harvard's Mark II computer, have seen the transformation of computer technology from rare and expensive machines operated by an elite group of extensively trained technicians and scientists to consumer products that are regarded by many as basic household appliances. The result is a techno-society in which computers and all other things digital and electronic are viewed as having value in and of themselves. Previous generations of consumer products, no matter how effective or useful they may be, are viewed as primitive and undesirable simply because they are not digital.

TECHNOSOCIETY AND LIBRARIES

Libraries and library employees have been deeply affected by this pervasive computer-centered technorevolution. Libraries were once the domain of highly sophisticated manual processes that were designed, developed, implemented, and refined by and for librarians. Efforts to teach library use to patrons notwith-standing, the greatest achievements of manual library technology – card catalogs, Cardex files, book charging systems, shelf arrangement schemes – served primarily to allow library employees to manage library operations and secondarily to make it possible for library employees to assist patrons in using the library. The essential functions of the manual library were performed behind the scenes by people specially trained in processes and techniques that for the most part had no analogs outside library practice.

The introduction of computers into library operations was at first relatively subtle and mostly concerned with improving behind-the-scenes functions. Computer-generated cards replaced typed cards in card catalogs, computer print-outs of serials holdings supplanted card files, 80-column punched cards appeared in book pockets, remote searches of offline databases could be requested. Few of these developments had a profound impact on library patrons or on library employees whose major tasks involved working directly with library patrons. Library employees whose jobs involved working directly with computers found their worklife transformed in more fundamental ways. New tasks had to be learned and new approaches had to be taken to preexisting tasks. In some cases, elements of the sophisticated functionality of manual library systems were lost in the transition to automated systems, many of which were adapted from systems originally designed for nonlibrary purposes.

The second generation of automated library systems, characterized by public access online catalogs and sophisticated online retrieval systems such as Dialog and Lexis-Nexis, had a much more obvious effect on library patrons and public service staff. The changes in processes, procedures, and basic tasks that had previously affected technical services employees now found their way to the public arena. The demand from patrons for assistance in interpreting the OPAC intruded on the daily routine of the reference desk. The introduction of online retrieval systems, whether directly available to the public – or more frequently – mediated by a professional, required the acquisition and maintenance of new skills and techniques. The tasks of remaining current with changes in search processes and understanding new databases were in many cases as over-whelming as dealing with patron demand for online searches.

Microcomputers define the third generation of library automation. It would be difficult to overstate the role microcomputers have played in the transformation of libraries. These relatively inexpensive, easy to understand and use devices made it possible to extend automation to the smallest of libraries, changed the ways in which database services were delivered, and extended automation into formerly unaffected areas such as routine office and business operations. Libraries responded early to the public need for access to computer hardware and software by installing public access microcomputers. Microcomputers represented the democratization of computerization in libraries as well as in society at large.

The fourth and most recent generation of library automation is defined by the Internet and by the World Wide Web in particular. The Internet has done more to eliminate barriers between libraries and patrons than all the previous efforts to build library networks and consortia. The high public visibility of the Internet has greatly and rapidly intensified patron interest in access to computers in libraries, frequently for purposes that have no direct relationship to the tradi-tional and historical purposes of libraries. The wealth of information available on the Internet has expanded access for libraries of all types and sizes, making it possible to quickly and easily retrieve information that was once the sole province of large research libraries.

The introduction of successive waves of automation into the library environ-ment has provided a vast array of benefits and produced an unprecedented information resource base for libraries and their patrons. However, those benefits and advantages have not been attained without accompanying costs. The most profound of these costs may be the impact on those professionals who must function in an environment in which technology has changed not just the tools, but the entire structure of work and the most closely held sense of professional identity.

APPROACHES TO CHANGE

Many people clearly believe that the major source of stress in a highly tech-nological workplace is the technology itself. Things are actually not nearly that simple, and, in many cases, the technology serves more as a decoy than as a true source of stress. Technology is primarily a manifestation of change, not a source or cause of change. Change, then is the most significant source of stress in an electronic environment. Understanding how people approach and respond to change is a major factor in treating the stresses associated with changing technology.

According to Flower (1996), there are three basic human approaches to change: avoidance, acceptance, and dominance. Each of these approaches derives from an unstated, frequently unrecognized personal goal that influences each individual's attitudes toward and responses to situations in which change appears likely or inevitable.

Avoidance

Avoidance behavior is based on an internal belief that change is unacceptable, that the impending change cannot and will not affect the individual in ques-tion, even though it may affect others. Avoidance is typically associated with a strong sense of identity with the status quo. When faced with an administra-tive decision to change the name "Reference Department" to "Information and Research Center", for instance, the reaction may be "I'm a reference librarian, I work in a reference department, and reference is what I do. No matter what they call it, I'm going to continue to do what I have always done". The tried and true "we've always done it this way" lament is a cornerstone of avoiding change.

The appeal of avoidance is that it "actually works – but only for matters that turn out to be irrelevant" (Flower, 1996). The greatest problem with avoidance is that it may initially appear to be an effective means of dealing with impor-tant matters but prove over time to be ineffective. In such circumstances, the individual who has chosen avoidance as a response to change may find that he or she has become marginalized. As the climate or state of change progresses, the distance between the organization and the individual increases to the point that the individual is irrelevant to the endeavor that was the subject of change. In the case of the reference librarian who does not want to be known as an "Information and Research Advisor", avoidance of change may lead to ineffective performance of assigned responsibilities, isolation from other librarians, disciplinary action, failure to gain promotion, and professional stagnation.

Technoanxiety is an extreme manifestation of avoidance behavior. Technoanxiety is "the primary symptom of those who are ambivalent, reluctant, or fearful of computers" (Brod, 1984, 16). Technoanxiety may result in resistance to learning technology-dependent tasks. Although some individuals may overtly refuse to learn and use technology, it is more frequently the case that the technology-oriented task is allowed to sink to the bottom of the worker's list of priorities and never rises to the top. As jobs become more and more technology-centered, a worker's ability to simply shove the undesired task aside decreases, often resulting in increased technoanxiety. Technoanxiety can produce physical symptoms such as headaches, irritability, and insomnia. These symptoms are very real. Arnetz and Wiholm (1997) found significantly elevated levels of stress-sensitive hormones in employees in technology-centered jobs who scored high on a general occupational stress scale.

In many cases, technoanxiety does not become apparent to anyone other than the individual experiencing it until it has reached the acute stage. Technoanxiety is rarely the result of a conscious decision to avoid using the technology. The employee wants to fit in, wants to do a good job, and wants to learn the technology, but is overwhelmed by the task of doing so. Technoanxiety tends to build into a cycle of failure: the worker tries to use the technology, inevitably encounters a problem or produces an error, and incorporates that encounter into the structure of the anxiety. The next problem or error seems larger and more serious and produces even more anxiety. A person suffering from technoanxiety is especially prone to sensing an acceleration of time and feels that events run together in a blur (Brod, 1984). Eventually, the state of anxiety may become so intense that the employee gives up all together. This sometimes results in the abrupt resignation of an employee who had shown no outward signs of job-related stress. Because technoanxiety so frequently is invisible until it becomes critical, it is essential to develop management strategies that make it possible to identify employees who are at risk.

Acceptance

Some individuals respond to change by adopting a stance of "whatever happens is all right with me". Although acceptance is essential to effectively coping with change, an excessive or insincere posture of acceptance is harmful both to the individual and to the shared endeavor. Over-eagerness to accept change may be a form of passive-aggressive behavior in which the individual is in an unconscious state of submissive resentment. People who are overly ready to accept change "are the easy-going, agreeable people who suddenly walk out on the marriage, chuck the job, sour on the project" (Flower, 1996). A librarian who agrees to all changes equally and without analysis has abdicated professional responsibility and is unlikely to be useful as a team member.

Dominance

For some people, the goal is to win. This can manifest itself in many ways, all of which have to do with identifying and defeating an opponent. In a very real sense, change itself is always the opponent, although the individual who responds to change with a desire to dominate typically has a much more immediate and specific opponent in mind. If the department head has introduced a new approach to documenting success in answering reference questions, the librarian who approaches change through domination may respond by engaging in activities designed to demonstrate that the new approach is a bad idea. If it can be demonstrated that the new approach is too difficult to understand, impossible to apply consistently, consumes too much staff time, or yields flawed results, he or she has won and the department head has lost.

The fatal flaw in dominance as an approach to change is that the dominance model requires that there be a winner and a loser. Competition in the dominance model is not an approach to reaching a desired goal, it is the goal. In some individuals, dominance goes beyond competition and manifests itself as technoobsession. Technoobsession is more insidious than technoanxiety in that the symptoms of obsession superficially seem to be positive behaviors. People suffering from technoobsession "push themselves in a constant effort to improve their work performance" (Brod, 1984, 92). These people learn new systems and routines readily and frequently develop their own shortcut approaches to accomplishing tasks. The volume of their output is impressive and they are frequently perfectionists who produce almost no errors. To the extent that the goal is output, technoobsessed employees appear to be ideal workers.

The invisible problem is that technoobsessed employees work only within the limits imposed by the technology. They embrace the technology to the point that they begin to identify more with the technology than with the library, their coworkers, or even their personal lives. "They don't take breaks, they don't talk about non-work subjects, they don't think abstractly, and above all they don't question the reason for doing the job they're doing" (Brod, 1984, 92). Technoobsessed individuals tend to not only take work home with them, but to choose to engage in activities outside of work that are structurally similar to the tasks performed at work.

Technoobsession results in rejection of tasks that are not related to the technology. Although the employee may have a range of tasks to perform, he or she will tend to focus on those that are gratifying to the obsession and simply ignore those that are not. Unlike employees suffering from technoanxiety, who tend to feel that time moves so rapidly it is impossible to keep up, employees suffering from technoobsession get lost in their tasks and don't sense the passage of time at all (Brod, 1984, 93). When asked by a supervisor why an alternative task was not completed, the techno-obsessed employee will quite honestly answer that he or she "just didn't get to it" and may point to the large amount of output generated by the task that is the focus of the obsession.

Benefitting from Change

The ultimate goal in responding to change is understanding how the change will be beneficial and adding value to the benefit inherent in the change. Avoidance, acceptance, and dominance are all tactical approaches that have to do with surviving change. The real goal is to thrive in the midst of change, to take the challenges and risks of change and turn them to a recognizable benefit.

The notion that every event, positive or negative, is a learning experience is trite, hackneyed and almost always true – if the individual to whom the positive or negative occurrence has happened is determined to learn from it. The paramount problem of avoidance, acceptance, and domination as responses to change is that extreme adherence to any is unlikely to lead to any meaningful form of learning.

The healthy response to change is a mixture of avoidance, acceptance, and domination. The librarian's professional responsibility is to recognize tendencies toward each and to utilize the positive aspects of all three. This means consciously adopting the critical view of avoidance, exploiting the positive philosophy of accepting change when it must take place, and occasional venturing into planned obsession when it is necessary to learn something in an intense manner.

COMPONENTS OF TECHNOSTRESS

Kupersmith (1992) has identified four components of technostress: performance anxiety, information overload, role conflicts, and burnout. These components apply most obviously to individuals who are experiencing technoanxiety, but may also apply to those who are subject to technoobsession. Technoobsessed people, for instance, often experience extreme performance anxiety and are compelled to achieve levels of performance greatly beyond those actually required for successful completion of the tasks for which they are responsible. Individuals who exhibit the characteristics of technoobsession are frequently candidates for extreme burnout.

Performance Anxiety

The first component of technostress, performance anxiety, has to do with fear of failure. As an individual's self-perceived ability to cope with technological change decreases, a vision of impending doom emerges. There is a growing conviction that it is only a matter of time before performance failures will be apparent to others, and catastrophe will surely follow. Lowenthal (1990) studied the relationship between reference morale and reference performance and found that "perceived tension, stress, and strain appeared to be very closely related to success of performance" (pp. 380–393). He also found that high levels of anxiety, including "loss of self-confidence and general manifestations of anxiety" were associated with performance failures. The relationship between performance anxiety and actual performance, then, is one of a declining spiral in which anxiety impedes performance and reduced performance heightens anxiety.

The Pace of Change
Inability to keep up with the pace of change is a common complaint in the expanding world of librarianship. Rose, Stoklosa and Gray (1998) identified keeping pace with advances, such as new software, hardware upgrades, and system enhancements, coupled with insufficient or ineffective training, as major sources of stress at the reference desk. The need to constantly learn and relearn electronic reference resources, combined with the accelerated pace experienced by many reference librarians in recent years, creates an environment in which professional competence is continually challenged and tested.

The Pace of the Job
Another typical source of workplace stress is the pace of the job. A job is least stressful when the worker can select the pace and adhere to it. Although there are natural variations in pacing both between workers and within the work of an individual, everyone has an optimal pace and works best when the work matches that pace. Routinization and pacing work together to create a working rhythm.

One of the effects of computer technology is a tendency for the system to influence the pace. The worker may begin to feel that it is necessary to work as rapidly as the system will allow, even when there is no real benefit to operating beyond the worker's normal pace. The worker experiences the stress of being constantly rushed by the system and may experience a constant feeling of being behind, of never being able to accomplish all that should be done. In other cases, the pace of the system may be slower than the optimal pace of the worker, either due to features inherent to the system or to a technological problem such as a saturated network. When the system is slow, the worker becomes impatient and frustrated and feels that his or her time is being wasted. Experienced librarians accustomed to answering questions and thereby completing specific, discrete transactions now must engage in teaching, which requires more time, leading to another stressor – inadequate staffing.

Technologization and Loss of Control
The introduction of successive waves of automation into the library environment has provided a vast array of benefits and produced an unprecedented information resource base for libraries and their patrons. Those benefits and advantages have not been attained without accompanying tangible costs. One of the most significant costs is lack of control.

Tools
Libraries have moved from a time when essential processes and tools were designed explicitly for the unique library endeavor to an era in which many of the most important processes and tools were created with no reference to the library. Many of these new processes and tools are demonstrably inferior to those that the library professions spent decades developing and perfecting. The classification schemes employed by Internet clearinghouses such as Yahoo! directly violate many of the basic principles of classification and search protocols that have underlain library practice. World Wide Web search engines represent a startlingly primitive approach to information retrieval when compared to the sophistication of more established systems such as Dialog. Most Web-based OPACs are substantially less capable than the stand-alone or Telnet-accessible systems that preceded them.

Individuality
Individuals also experience loss of control in a highly technological environment. Manual systems typically allow for variations in the ways in which tasks are performed and can readily be tailored to the preferences or needs of the individual worker. Even when there is one optimal, desired, or officially sanctioned way to use a manual system, there is generally some margin for variation. Highly technological systems are usually designed with one correct approach

to the task in mind. Failure to work according to the design of the system may result in inefficiency or error.

Although the worker has some ability to tailor the hardware environment by positioning the monitor, selecting an ergonomically sound keyboard, or choosing a trackball rather than a mouse, there is little the user can do to influence the software environment. There are few computer systems that adapt themselves to the preference or style of the user. Instead, the user must adapt to the system. The loss of control is exacerbated by the lack of meaningful standardization across systems. Moving from one software system to another, or even from an older release to a newer one, can be an incredibly frustrating experience as the user finds that familiar processes no longer apply and new functions must be learned.

Predictability

For many people, work is an inherently stressful endeavor. Every workplace is or can be a source of multiple stressors. One of the strategies individuals use to offset stress is routinization – the transformation of tasks into predictable, repetitive processes. Making work routine reduces the elements of surprise and unpredictability that are major sources of workplace stress. Although excessive routine may lead to boredom, which is itself a source of stress, the ability to forecast work situations is important to most workers.

Working in a highly technological environment becomes stressful in part because of the reduction of opportunities for routinization. An unfortunate aspect of technological solutions to workplace needs is that technology fails – the photocopier jams, the printer runs out of toner, the operating system freezes, the network crashes, the remote site is unavailable. Although technological failures are generally an essentially random occurrence, their impact on worker perceptions is frequently that the technology always fails when it is most needed. Manual systems may also fail – the almanac hasn't been reshelved properly, someone removed the card from the catalog, the needed issue has been sent to the bindery – but those failures are much less conspicuous and less traumatic than technological failures. More importantly, manual failures can more often than not be addressed easily – the almanac probably isn't the only source of the answer and there are other entries in the card catalog. When the network crashes, though, access to networked information becomes an impossibility.

These technological failures not only break the routine, but make it difficult to establish any sense of predictability in the workplace. The loss of routine that accompanies technological failure is joined by other interruptions, such as an increased volume of telephone requests and the expanding need to address patron education on demand. It creates a working environment that is dangerously unpredictable. At its extreme, the stress of loss of routinization creates a situation in which the worker is reluctant to begin tasks as a result of fear that the task will be interrupted by a technological failure. In some cases, the worker may elect to use a less appropriate manual approach to the task as a means of avoiding technological failure.

Information Overload

Information overload is "a state occurring when the amount or intensity of environmental stimuli that needs to be processed exceeds a person's stress and capacity, potentially leading to disregard of some important information" (Corsini, 1999, 486). The concept of information overload has been discussed for more than a century, although the term is of more recent origin. Otlet and La Fontaine's interest in universal bibliographic control (1895) and Wells' notion of the world brain (1938) were early manifestations of concern for information overload. The concept was revisited by Bush in his discussion of the Memex (1945).

Although frequently presented in terms of the overabundance of documentary, broadcast, and digital information that has arisen in the post-World War II era, information overload is actually a more complex phenomenon. True information overload involves any excess of stimuli that interferes with human information processing. Some common symptoms or warning signals of information overload include spending too much time dealing with trivia, inadvertently ignoring essential information, failure to recognize the importance of information, responding to information without analyzing the information, and assigning arbitrary value to information (Alesandrini, 1992).

Klapp (1986) discussed two fundamental elements of information overload: noise and banality. Much of the information barrage to which modern society is subjected is essentially passive: rather than consciously seeking information, people are constantly presented with information that requires no active role for the recipient. In many cases, the recipient cannot turn the stimulus off and has no choice in selecting stimuli. The information being received, some of which may in fact be useful, takes on the characteristics of noise. Inputs lose meaning and it is frequently difficult to distinguish between foreground and background stimuli. In extreme cases, excessive noise in the environment produces a form of dyslexia in which an individual loses the capacity to successfully interpret information in any form. At the same time, much of the information to which people are subjected is "sterile and redundant", resulting in "banalization" (Klapp, 1986, 2). The combination of noise and banalization results in degradation of the value and impact of information.

One of the conundrums of information overload is that the stress of attempting to process excessive stimuli frequently leads to a state of boredom. The repetitive nature of many jobs couples with the inability to process inputs and produces a situation in which the employee is no longer engaged in the tasks he or she does, but becomes a kind of robot, delivering services without reflecting on them or benefitting from the experience. This boredom frequently spills into the individual's personal life as well, producing a situation in which he or she derives little or no satisfaction or pleasure from either the working environment or home life. Although social scientists have historically described boredom in terms of lack of stimulus, there is a growing sense that boredom is the result of lack of *meaning* and that any volume of stimulus that is devoid of meaning can result in boredom (Barbalet, 1999).

Role Conflicts

A librarian plays many roles: selector and filter of collections, organizer of information, interpreter of policies and procedures, retriever of answers, research and reading adviser, teacher, manager, and more. The advent of ready direct access to information via the World Wide Web has produced a situation in which patrons increasingly believe themselves to be both responsible for and capable of managing their own information needs. In many libraries, the role of the reference librarian has undergone a major shift in which the direct answering of questions has declined significantly but demand to teach use of the World Wide Web and interpret the results of patron searches has increased dramatically.

Threats to Professional Competence

One of the normal sources of stress in a public service occupation is the desire to do a good job – to serve the patron effectively, efficiently, and appropriately. The ability to do a good job is accentuated when the worker is confident of his or her knowledge and skills. The lack of control that accompanies technological solutions to workplace needs acts to undermine that confidence.

According to McGrath (1999), uncertainty regarding professional competence is manifested in four basic fears:

(1) "I can't make this thing work".
(2) "It's my fault".
(3) "The machine doesn't like me".
(4) "There's never enough time anymore".

The general lack of control inherent in computer software leads to the first two fears. Even highly experienced users of software encounter circumstances in which it is difficult or even impossible to determine how to accomplish a specific task. When a user of a computer system can't determine how to "make this thing work" he or she tends to become lost in a cycle of increasingly frustrating, increasingly incorrect attempts to solve the problem. There is a natural tendency to engage in self-accusations of incompetence and guilt. The user adopts the belief that he or she should know how to make the system work and is at fault for not knowing. Ultimately, many people assign anthropomorphic characteristics to the computer, concluding that the computer itself is preventing the accomplishment of the task. This may be accompanied by a very real desire to physically assault the computer and may evolve into actual acts of violence and destruction. A 1999 study by Concord Communications found that more than 80% of network managers polled reported instances of deliberately destructive behavior inspired by computer-related frustration (Armour, 1999).

Loss of control, collective or individual, leads to a feeling of lost ownership. The individual can no longer utilize the traditional strategies that serve to make the job uniquely and comfortably individual. There is a sense that the basic functions and processes that have defined librarianship are no longer the property of librarians.

Threats to Professional Identity
In addition to threatening professional competence, the advancing digitization of the information environment introduces threats to the core identity of the library professions. Externally, there is a fairly pervasive anticipation that the virtual library will entirely supplant the "traditional" library and that "in lieu of librarians we will have programmers and database experts" (Churbuck, 1993, 204). This expectation is founded in the false beliefs that the entirety of the human record will soon be available (or already is available) on the World Wide Web and that the management of digital information doesn't require the skills and abilities of librarians. Although these beliefs seem ridiculous to the knowledgeable, they are both acceptable and appealing to the uninitiated. When adopted by people in power, such as library board members or university administrators, these beliefs have the potential for becoming extremely damaging to the library endeavor and may become direct threats not only to professional identity but to continued employment. When carried to their logical conclusion, these false beliefs may lead to reductions in library resources at a time when there is a need to expand the library's role in meeting the needs of distance learners, supporting new outreach programs, and serving new populations.

Internally, the impact of technology in libraries may be viewed in terms of the deskilling of the professional. Time that could be spent in professional activities is instead spent clearing paper jams and changing ink cartridges.

Instead of answering reference questions, staff members find themselves explaining the workings of e-mail. Every moment spent dealing with the superficial workings of the technology is a moment not spent in the kind of direct patron service that makes librarianship an attractive profession.

Some librarians object to the increased need to teach patrons how to make effective use of information resources. To a librarian whose education and career have focused on question-answering, the shift to patron education may seem to be primarily a distraction from the main purpose of reference service. Librarians who do not have backgrounds in education may find not only their identity but also their competence challenged by the expanding role of the reference librarian as teacher. Van Fleet and Wallace (1999a) found within one staff cohort group radically conflicting views of the role of the librarian as educator: while some reference staff members felt that acting in an instructional role was definitely deprofessionalizing, other staff members felt that their involvement in patron education added to their professional credibility.

Furthermore, some members of the profession have seemingly adopted the public view that working with digital resources doesn't require professional abilities. It has been suggested on the floor of the American Library Association Council that librarians can learn effective use of the World Wide Web by observing teenagers. When such assertions come from the general public they are disturbing; when they come from within the profession, they are alarming.

Patron Perceptions and Patron Education
Perhaps the most comprehensive sources of stress in the digital electronic reference environment are the expectations and needs of library patrons. The overselling and overstating of the World Wide Web that have taken place in the popular media have greatly heightened patron perceptions of what digital media can deliver. A common patron perception is that the library has been magically transformed from a passive source of printed materials to a comprehensive electronic source of all knowledge and information. Patrons actually reject readily available information in print-on-paper form in the belief that information provided electronically must be more timely, more comprehensive, or in some other sense better. There is a sense that every answer to every question must be available in digital form. Rose, Stoklosa and Gray (1998) term this over-acceptance of digital formats *technological idolatry*. More than one librarian has experienced the phenomenon of the classroom teacher who not only encourages students to use the Web to find support for assignments but actually prohibits the use of non-Web resources.

The "friendliness" of the World Wide Web has been so exaggerated that many patrons expect to find precisely and exclusively what they want with no analysis of the information need and no assistance from a staff member – and to find it on the first attempt. Patron frustration when exactly the right information does not immediately appear translates into anger and resentment, often addressed directly to the library staff member whose help was not initially desired or sought. One librarian noted: "We don't get the luxury of the easy question any more. We get the hard ones, after they've tried and failed. And then they're frustrated with the system already and pass that along to us" (Van Fleet, 1999). Other patrons are excessively and uncritically pleased that they have found information on the Web, even when they are ill-prepared to assess the quality of the information found. It is greatly frustrating to librarians to know that patrons are relying on poor – sometimes dangerous – sources of information from the Web when reliable information is available in other formats.

A disturbing trend that seems to be at least in part related to increased dependence on high technology is a general loss of social civility. The media are full of stories of road rage, air rage, and other instances of antisocial behavior sparked by seemingly trivial occurrences. The perceived need for a specialized system of "netiquette" distinct from ordinary and traditional rules of etiquette is evidence that there is a gap in the congeniality of daily human interaction. A cartoon in the *New Yorker* (Steiner, 1993) pictured a dog at a computer with the caption, "On the Internet, no one knows you're a dog". Unfortunately, many users of the Internet give the impression that they don't really care that the entities with whom they are interacting are human beings. Whether the e-mail phenomenon of flaming leads to or is caused by a general lack of respect for others, there is a definite trend toward treating people as if they are machines or – worse – as if they are automatically adversaries. Impatience is rampant and frequently is only a precursor to acts of outrageous anger sparked by trivial events.

Burnout

Burnout is "a syndrome of emotional exhaustion, depersonalization, and reduced personal accomplishment that can occur among individuals who do 'people work' of some kind" (Maslach, 1982). Gorkin (1998) describes four *stages* of burnout: exhaustion; shame and doubt; cynicism and callousness; and failure, helplessness, and crisis.

Mental, and Emotional Exhaustion
In this stage, the individual is still coping, but is beginning to function less well than in the past. The major symptom is simply not having as much energy as in the past. People who are accustomed to having very active lives outside of the workplace find themselves going home from work, scrounging a meal

rather than preparing one, and collapsing in front of the television until bedtime. They often take work home but rarely actually do work at home. Basic domestic tasks go undone because there just isn't enough psychic and physical energy to do them. At work, things begin to slip. Tasks that were once completed automatically but well now require an unusual amount of concentration and still don't come out well. There is a feeling of boredom and discontent, even when doing things that were once exciting and energizing. No matter how much work the individual does, there is frequently a feeling of spinning wheels, of taking excessive time to produce less real work. Concentration fades and the employee becomes increasingly conscious of the passage of time during which he or she feels that no work is being accomplished. The lack of sufficient energy to do things well leads to conscious shortcuts; the individual is quite aware that these alternative approaches are not the best way to complete the task, but feels that the job won't get done at all if he or she has to invest the energy necessary to doing it right. There is usually a very great awareness that it used to be possible to do things right and that something very important is missing.

Shame and Doubt

The awareness that all is not well and that tasks are not being done as effectively or as efficiently as used to be the case leads to feelings of guilt and shame and a growing sense of self-doubt. New responsibilities, no matter how intrinsically attractive, are assumed with reluctance and trepidation. The feeling that current responsibilities are not being met pervades the employee's approach to, and attitude toward, work. Confidence in past performance is eroded as well. Where once there was valid knowledge of abilities, skills, and achievements, now there is a growing conviction that it was all a sham, that the employee never actually did a good job and now the truth is coming out. Self-esteem hits an all-time low. A sense of loss emerges and the employee begins to feel deeply vulnerable.

Cynicism and Callousness

The third stage of burnout is a defense against the sense of loss and vulnerability that results from the shame of not doing well and the fear that the employee's past positive performance is all an illusion. The shame and doubt don't matter as much if the employee can believe that nothing really matters anyway, that the basic endeavor of the library is trivial and unimportant, and that patrons aren't really in need and don't benefit from the library's services. Patrons receive terse, unhelpful replies to their requests for service. They are made to feel that their questions are worse than foolish, that they are not really worthy of the librarian's attention, and that there is something much more important that the librarian should be doing. This is the stage that most people associate with burnout – the individual who has lost interest in doing the job and is just marking time, waiting for something to happen that will put an end to his or her misery. In the meantime, everyone else is expected to share in that misery. Instead, other employees choose to put as much distance as possible between themselves and the employee in the third stage of burnout. That distancing, of course, both justifies and exacerbates the cynicism and callousness.

Failure, Helplessness, and Crisis
In the fourth stage of burnout, the individual begins to feel completely unable to cope. He or she becomes excessively thin-skinned and worn down. Coping strategies that once worked – even those of cynicism and callousness – are no longer effective. Reactions to minor negative events have a major, irrational impact from which it is nearly impossible to recover. The urge to simply quit, to walk away from it all, begins to dominate and there may even be suicidal tendencies. The employee withdraws from interpersonal contact and becomes increasingly introverted. Ultimately, he or she just can't face the day. Some people become physically ill at the thought of going to work; others stay in bed and call in sick, even though they don't really feel physically ill.

A study of the burnout phenomenon among reference librarians in medium-sized to large public libraries found that about one-fifth of respondents scored in the high burnout category, while only one-fifth "indicated a strong positive feeling of personal accomplishment" (Smith, Birch & Marchant, 1984, 83). The age of this study suggests that burnout in the reference environment is not a new phenomenon and that reference work is an aspect of the library and information professions that is particularly vulnerable to burnout. It seems probable that the factors observed by Smith, Birch and Marchant still prevail and have been exacerbated by the expanding role of technology in the worklife of reference staff.

TECHNOSTRESS ASSESSMENT

There appear to be no standard tests for assessing technostress per se, although there are a number of related instruments. The Technophobia Measurement Package (TechnoStress Measurement Tools n.d.) – which consists of the Computer Anxiety Rating Scale (CARS-C), the Computer Thoughts Survey (CTS-C), and the General Attitude Toward Computers Scale (GATCS-C) – has been widely administered and subjected to research testing. The Maslach Burnout Inventory (Maslach, Jackson & Leiter, 1997) has been extensively used and thoroughly validated, but focuses on a single factor in the technostress equation. The Moch Stress Curve provides an approach to diagnosing the stages of burnout, but also focuses on burnout rather than technostress (Moch, 1998). A number of tests, such as Shrink's "burnout inventory" (1996) and Potter's "Burnout Potential Inventory", (1993) can be found online.

The Reference Stress Inventory (Appendix) was developed by Van Fleet and Wallace (1999b) for use as part of a series of workshops designed to help reference staff members identify and cope with technostress. Although the Reference Stress Inventory has not been formally validated, it was used effectively with two groups of professional and paraprofessional reference staff members to provide a focus for discussion of personal experiences with technostress in the reference environment.

PERSONAL COPING STRATEGIES

Coping strategies can be divided into two basic categories: emotion-focused strategies and problem-focused strategies.

> Problem-focused coping refers to efforts to improve the troubled person-environment relationship by changing things, for instance, by seeking information about what to do, by holding back from impulsive and premature actions, and by confronting the person or persons responsible for one's difficulty . . . Emotion-focused (or palliative) coping refers to thoughts or actions whose goal is to relieve the emotional impact of stress. These are apt to be mainly palliative in the sense that such strategies of coping do not actually alter the threatening or damaging conditions but make the person feel better. Examples are avoiding thinking about the trouble, denying that anything is wrong, distancing or detaching oneself as in joking about what makes one feel distressed, or taking tranquilizers or attempting to relax (Monat & Lazarus, 1991, 6).

Despite Monat and Lazarus's implication that problem-focused strategies are of greater and more direct value than emotion-focused strategies, Hudiburg (1996) has suggested that the two categories are interrelated and that "what may be an optimal strategy is highly dependent upon the individual's situation and perception."

The approaches discussed below are suggestive rather than prescriptive. There is no single recipe for effective management of any kind of stress and what works for one individual will not work for another. An essential step in developing a strategy for coping with technostress is determining whether it is possible to employ a preventive, self-directed approach or to seek professional assistance.

The best approach to technostress, as to all health issues, is prevention. Kupersmith (1992) suggested the following basic categorization of strategies for coping with technostress and preventing burnout: relaxation and health, cultivating a positive attitude, time management, goal setting, and exploiting learning opportunities.

Relaxation and Health

Although general relaxation, health, and constructive leisure activities clearly fall into the category Monat and Lazarus refer to as "palliative", there can be no question that a generally healthy individual is better prepared to cope with change and stress than a less healthy person. Stress induces physical changes in heart rate, blood pressure, central nervous system activity, and other areas. Everly (1989) has described a "relaxation response" that is the opposite of the stress response and that can be induced consciously to counter stress. Effective techniques for inducing the relaxation response include meditation, yoga, and laughter.

Effectively coping with technology-induced stress requires recognition of the potential and real conflicts between a technology-rich life and a healthy life (Baldwin, 1985). Some tactics for living a healthier life in a technocentric environment include engaging in leisure activities that are undemanding, slow-paced, and non-goal-oriented; avoiding non-work-related use of information technology; and emphasizing outdoor recreational activities. Exercise has been shown to have a direct, positive impact on work-related stress (Long & Flood, 1993). The basic goal is to find an appropriate balance of the beneficial use of technology in the workplace and a mentally and emotionally satisfying personal existence (Baldwin, 1985).

Some specific tactics for maintaining a healthy lifestyle in a technology-dominated environment include taking true breaks, not working breaks or desktop lunches; developing the habit of getting away from the desk and walking, whether outside or in the building; and learning to breathe effectively, as is taught in yoga and relaxation classes.

Attitude

This is also a palliative according to Monat and Lazarus's classification. Cultivating a positive attitude is sometimes a difficult task and may be impossible for an individual who has reached an advanced state of burnout. One approach to cultivating a positive attitude is what Meichenbaum termed *stress inoculation*. Stress inoculation training is a three-part process in which a professional: (1) works with a patient to learn his or her personal response to stress, (2) rehearses stress-inducing scenarios and prepares strategies for addressing them, and (3) applies those strategies to combat stress. Stress inoculation training has been successfully used in group settings to improve occupational health (Bamberg & Busch, 1996) and as a means of reducing burnout (Freedy & Hobfoll, 1994).

Part of cultivating a positive attitude comes from appreciating one's own perspective. The reference librarian has a privileged position in being able to observe first hand the impact of library policies, procedures, and applications. The reference librarian also has an obligation to ensure that those who are responsible for formulating those policies and allocating resources to carry them out are provided with the front line librarian's expertise and experience. Cultivating a positive attitude includes approaching problems, including technology-related ones, in a professional manner. Rather than feeling overwhelmed, angry, or frustrated, the effective reference librarian engages in a positive process: identifying problems; documenting problems; and reporting problems and potential solutions management in an objective and straightforward manner. If administration does not respond in kind, the problem is not with technology.

Time Management

The overabundance of popular treatises on how to manage one's personal and/or professional time is more than ample evidence of the importance of time management. Time management techniques are commonly presented as cures for the problems of information overload. Alesandrini (1992) discussed many of the most prevalent time management techniques and dismissed them as "myths" that set unattainable goals and focus excessive attention on trivial details such as never handling any piece of paper more than once. According to Alesandrini, "traditional time management tools such as time planners and calendars can undermine effectiveness because of their emphasis on daily planning" (34). Alesandrini emphasizes "seeing the big picture" – expanding one's perspective rather than excessively focusing it and paying constant attention to contextualization.

Koch (1998) has applied the 80/20 Principle, which has been used to describe the relationship between collection size and demand in libraries, to issues of time management and productivity. Koch, like Alesandrini, contends that most time management techniques are ineffective and counterproductive. Koch's extension of the 80/20 Principle states that 80% of results, whether personal or shared, are generated by 20% of effort expended. The challenge, then, is to identify and isolate the 20% of effort that is most productive and maximize its effectiveness.

Despite the concerns of Alesandrini, Koch, and others, time management is a very big business and many individuals have benefitted substantially from basic time management tools and techniques. What can be learned from Alesandrini and Koch is that time management must have a purpose – time management for the sake of time management is a futile and wasteful endeavor. The fundamental question an individual must ask is "how do I want to use my time and why do I want to use it in that manner?" The use of the word *want* is deliberate – although time on the job is often a function of demands imposed from outside the individual, those demands will turn into productive time management only if internalized and turned into tasks the individual wants to achieve.

Goal Setting

Perhaps the most insidious aspect of burnout is the loss of a sense of direction and purpose. The combination of information overload and increasing workplace demands tend to induce fears of failing professional competence. The individual begins to think "I should be able to keep up, I used to be able to keep up, something's wrong with me". It is essential to recognize that already busy people cannot take on new tasks without giving up preexisting responsibilities. Setting realistic, attainable goals is a key element in avoiding burnout and minimizing workplace stress. This is not solely the responsibility of the individual, but the individual must accept and assume some responsibility for working within the framework of goals that are personally understood and acceptable and that can be realized.

Goal setting is not in and of itself a solution. Many librarians have experienced the inherent futility of structured strategic planning initiatives in which planning itself appears to be the ultimate purpose rather than attainment of the goals that underlie the plan. Goal setting without assessment of the achievement of goals is an empty exercise. Furthermore, there must be active recognition that goals must be susceptible to evolutionary change. It is the norm rather than an exception that working toward achievement of a goal has the effect of redefining and reshaping the goal.

Learning Opportunities

"Sociologists recognize the impact of frequent, inescapable change on both institutions and individuals, and find that an ability to learn and adapt throughout the lifespan are important in coping with this change ... Psychologists recognize the growing sense of alienation and helplessness in the face of this change and the need for an effective coping mechanism ... Businesses have recognized the challenge of the world economy and the need to develop human resources through continued education" (Van Fleet, 1990, 197). Baldwin (1985) has pointed out that technology affects human functionality in a variety of ways, including highlighting the need for constant learning and change, producing an environment in which it is truly impossible for any professional to maintain a sense of confidence in the extent to which he or she is aware of current events, developments, or practice in the profession. In a time of accelerating technological transition, taking advantage of opportunities to learn is a responsibility that is shirked at the risk of loss of professional competency and professional credibility. There is a need for every library employee to become an active learner and an advocate for continuing professional education. According to Hudiburg (1996), "the first and foremost way to cope is through education and training" (a problem-focused coping strategy).

MANAGEMENT STRATEGIES

Although individuals can do much to alleviate the impact of technostress, there is an overriding need to address technostress as a management challenge and to develop strategies for managing technostress at the organizational level. Unfortunately, much of the management literature related to the introduction of technology focuses on "suggestions for how to change the work organization [that] are rather categorical and based on a technocratic view of the way people organize and work together" (Arnetz & Wiholm, 1997, 35). There is a tendency in the management literature to blame the victim by ascribing technostress to failings within individual employees rather than recognizing technostress as a legitimate and serious management concern.

Seven major management strategies for helping staff cope with technostress include analysis of change, environmental scanning, providing support for individuals, fostering cooperation, managing the flow of information, emphasizing hands-on practice, and distributing expertise.

Analysis

Analysis has to do with the search for underlying causes. Change does not happen randomly. It is rarely the case that change is truly revolutionary. Most change is the result of processes that appear revolutionary at the time they are first experienced but can with proper analysis be seen to be evolutionary in nature. To many library patrons – and not a few library employees – the introduction of online public access catalogs seemed to be an overnight phenomenon that had no precedent and little explanation. The truth, of course, is that automation of library catalogs was an outgrowth of a pattern of changes in library operations with a history of several decades. Merely understanding that a specific change is less likely to be an isolated revolutionary event than to be a point on an evolutionary continuum is a major step in developing the analytical abilities and habits necessary to more effectively dealing with change.

Environmental Scanning

Analysis requires data. Developing a program of deliberate, continuous environmental scanning provides the data from which effective analyses can be derived. Most people maintain constant awareness of a very tightly circumscribed array of routine experiences – home, work, personal interests, the interests of friends. Even professionals tend to define their professional responsibility in terms of what will happen on the job today and tomorrow. What happens on the job the day after tomorrow may be influenced by factors outside the sphere of individual job-centered responsibility. Environmental scanning is the process of constantly looking for external, global factors that may lead to internal, local change.

Effective environmental scanning is truly continuous, not episodic. The concept of the New Year's Resolution loses all real meaning in a world in which a year encompasses dozens of significant changes. For environmental scanning and analysis to be useful and beneficial, they must be incorporated into the routine. When these processes are treated as an infrequently occurring special event, the results will be treated as special and incidental, not as integral and fundamental.

Ashley and Morrison (1997) emphasized the need to identify *signals* of change, such as demographic shifts, environmental trends, or other emerging issues that may serve as predictors of systemic or significant change. To the general principles of environmental scanning are added the need to challenge prevailing assumptions, the process of deliberately searching for ways in which change leads to vulnerability, and the construction of scenarios as an approach to preparing for the impacts of change.

Support for Individuals

Technostress frequently leads to feelings of isolation and disconnection from institutional or unit goals. Individuals lose track of their position in the shared endeavor and begin to believe that they are noncontributors. There is an intense need for managers to understand the talents, needs, and concerns of individual staff members and to ensure those support services that each individual needs. Kupersmith (1992) referred to the benefits of "management by wandering around". Many staff members may be reluctant to approach management, especially when they have real concerns about their own performance. In many cases those concerns are poorly-defined and the staff member feels responsible for producing a solution independently rather than by seeking assistance. Being on the floor and observing performance, engaging in informal conversations with staff members regarding the challenges of their jobs, and observing staff members interacting with each other and with patrons can provide the manager with critical information regarding how to provide appropriate support for individual staff members.

Cooperation

One of the essential management challenges in addressing technostress is developing an environment of cooperation and shared strategies. Individuals need to be brought into the processes of planning and decision-making, particularly in times of change. It is important to "remember that change is most successful when those who are affected are involved in the process" (Bennis, Benne & Chin, 1985, 226). There are many techniques for increasing involvement and fostering cooperation, one of which is the retreat. According to Kirkland and Dobb, "the retreat is one effective method which the administration and staff of an automating library may choose for responding to technological change" (1989, 505). A structured retreat creates a nonthreatening environment in which to explore key issues, establish shared viewpoints, and seek common solutions. Structure is an essential component in a successful retreat. A retreat must have a mutually understood purpose, must operate according to a carefully selected process, must lead to a tangible product, and must be people-centric. A critical element of a retreat as an approach to treating technostress is that the retreat must emphasize human values and interpersonal interaction.

The Phoenix Public Library exemplifies the value of creating opportunities for discussion and for creating an environment in which communications channels are open and in use at all levels. As part of a regularly scheduled series of professional development forums, Van Fleet and Wallace (1999a) applied a focus group-like approach to a staff development workshop. A significant number of staff members involved in the workshop felt that professional time was being wasted by the need to reboot public access computers; administrators responded by suggesting that a page could be employed for routine rebooting, thereby avoiding the need for reference staff to engage in such a low level activity. A major outcome of the workshop was an understanding that administrators do not automatically understand the concerns that are foremost in the minds of staff members and that sources of staff aggravation may have fairly simple solutions.

Flower (1996) discussed the importance of "abundant relationships". Most interpersonal relationships within institutions are narrowly proscribed. Meaningful relationships rarely link individuals across departmental or divisional boundaries. As a result, administrative units tend to become isolated boxes in which the range and depth of relationships are severely limited.

Similarly, the library as an administrative unit frequently has few or no meaningful relationships to other administrative units. The more complex and rigorously bureaucratized the institutional environment, the more restricted and focused relationships tend to be. Flexibility in coping with change requires an abundance of relationships that extend the reach and understanding of both individuals and units beyond bureaucratic borders. Extending relationships enriches the working environment and makes the workplace more pleasant. It also builds informal alliances that can become critical resources in times of change. The threat of change is greatly reduced when it is understood that change is systemic rather than localized. Such understanding can be facilitated by a broad network of relationships across departments and divisions of the organization.

Information Flow

One of the most productive roles of a manager is that of gatekeeper. The tendency toward individual information overload can in part be handled by establishing processes through which individual staff members are not solely and independently responsible for the tasks of identifying the information they need to do their jobs. Although it has been popular for several decades to refer to the information *explosion*, the problem is rarely a true overabundance of information. The real problem is usually an excess of *unorganized* information combined with a dearth of information available for effective use. Information in organizations tends to be rigorously compartmentalized on an unstated "need to know" basis. One of the benefits of an environment of abundant relationships is expansion of the availability of useful information, much of it embodied in individuals. Creating a situation in which useful information is abundant and available as a tool for managing change requires an administrative posture of openness and honesty – when information is withheld and secrets abound, the availability of useful information will be severely restricted.

There are many areas in which reference managers can act in a gatekeeper capacity. Providing syntheses and analyses of new library policies and procedures – not just those directly related to reference staff but also those that apply to all library employees – can ease the burden of learning and internalizing institutional changes. Front-line reference staff, who live intensely busy worklives, may find it difficult to maintain close familiarity with national trends; reference managers can work to proactively disseminate standards and guidelines such as those promulgated by the Reference and Adult Services Association or state library associations. An especially important management role is that of distributing and subcontracting gatekeeper responsibilities to other reference staff. A simple example of this process is that of asking staff members who attend conferences, workshops, or meetings to report what they learned or experienced at subsequent staff meetings.

Hands-On Practice

In the traditional environment of the now rather distant past, reference librarians may have had adequate time to examine reference works, assess their content and its value, explore the basic arrangement of the work, review the index, and generally internalize the role of the work within the overall reference collection. Additionally, print on paper reference sources are reasonably, although certainly not totally, amenable to rapid assimilation at the time of use. It was possible for a reference librarian to attain reasonable mastery of a large collection of print reference resources mostly through routine daily exposure to, and use of, the collection.

Electronic resources are much more difficult to learn via ad hoc processes. There is an intense need for hands-on practice, whether in the form of a structured workshop-style learning experience or as individualized trial and error. One of the most valuable things management can do for reference librarians is to recognize the need for hands-on practice and allocate time to staff to devote to hands-on exploration. Hands-on exploration should be an expectation, not a begrudged accommodation; evidence of time spent learning how systems work and how to exploit them to the benefit of patrons should be part of the evaluation process for reference staff.

Distributed Expertise and Authority

Hierarchical organizations in which decisions and accompanying actions are rigidly centralized are in a poor position for managing change. In an organizational setting that is optimized for change, every individual understands the realm within which he or she is empowered to make decisions and act on them. Risks and rewards are distributed among the individuals who are responsible for doing the work of the organization. The basic principle is to make every decision at the *lowest* level possible. The people who must implement and live with decisions frequently are the people who are best equipped to make those decisions.

A difficulty in distributing power is learning to live with the potential for failure – not everyone will make the right decision every time. Some individuals at lower levels will find it difficult to accept responsibility for making and implementing decisions. Fear of administrative retaliation or peer ridicule can be difficult to overcome. At the same time, some administrators find it very difficult to give up personal responsibility for decisions at all levels. Traditional organizational models suggest that top-level administration is ultimately responsible for everything, an attitude that encourages micromanagement and discourages the distribution of power. These are problems that must be overcome if the organization is to be capable of managed change.

Reference department managers can recognize and legitimize individual expertise by assigning to staff members responsibility for maintaining currency in specific subject areas or for learning and teaching particular tools. In some reference departments, day-to-day management activities such as reference collection development and management or desk duty scheduling is distributed among reference staff. Depending on local needs and local expertise, these assignments may be essentially permanent or may be reassigned on a scheduled rotation basis.

LEADERSHIP

The above strategies are based on the principle of managing the impact of change on reference staff. A further essential principle is that of providing leadership. Technostress is a prime example of the need to lead by example. A manager who is conspicuously a victim of technostress will be able to do little to alleviate technostress among reference staff members, even if he or she assiduously applies the seven technostress management strategies. Providing effective leadership requires building a team spirit in which the manager is sometimes player, sometimes coach, sometimes cheerleader and always closely connected to the team. The team effort can succeed only to the extent that team members believe in and support the goals of the team. Team members will not buy into a team effort in which the leader is subject to rules that differ from those of other team players. Although the seven management strategies for technostress can be applied in the absence of leadership and team-building, technostress reduction is a situation in which the results of management without leadership are likely to be shallow and ephemeral. In the worst case, there will be slight but noticeable false reduction of the symptoms of technostress among some individuals followed by a cataclysmic slide into an environment in which technostress is more pervasive and more invasive than before.

JOB DESIGN

Morris and Dyer described the role of *job design*: "A well-designed job is one factor influencing staff motivation, and well-motivated employees have the potential to contribute beyond the narrow specifications of their jobs" (1998,

283). Essential principles of good job design include variety, autonomy, feed-back, social contact, identity, responsibility and skill, achievement, opportunities to learn and develop, appropriate role load, minimized role conflict, minimized role ambiguity, and rest.

Variety

Although routinization is an essential function in assuring predictability and minimizing technostress, excessively routine tasks lead to boredom and fatigue. A well-designed job incorporates a mixture of tasks adequate to ensure that the staff member has a clearly-defined sense of variability and is not tied to a single repetitive task for an extensive period of time. A satisfying job requires a range of skills, not application of a single skill. Desk duty, the common denominator of reference work, is simultaneously the most directly rewarding and the most stressful of reference activities. As essential as desk duty is to reference, all reference and no alternative assignment is a fatal combination. Desk duty must be augmented and balanced by other assignments, such as reference collection development and management, design and development of pathfinders and other patron access tools, direct involvement in patron education and programming, or shared responsibility for department management.

Autonomy

A simple extension from the principle of task variety is the creation of an envi-ronment in which individual staff members justifiably feel that they are personally and individually responsible for determining what they do. Within the context of an established range of essential responsibilities, the employee should be able to determine the order in which tasks will be performed and should be evaluated in part on his or her ability to successfully build an indi-vidualized approach to accomplishing assigned duties. Although desk duty is typically dictated by a centralized schedule, flexibility in the scheduling of desk assignments should be a consideration. It may be desirable to deliberately rearrange desk assignments over time to avoid locking individual staff members into rigid work schedules. Assignments beyond desk duty, such as collection management or self-paced learning, usually don't need to be scheduled. In an optimal situation, the staff member is responsible for the timing and comple-tion of responsibilities that are not explicitly scheduled.

An annual or perhaps more frequent staff member-supervisor goal setting session can go a long way toward instilling a sense of self-direction and auton-omy. Such a meeting can provide the manager with an opportunity to reinforce the staff member's self-confidence in his or her range of responsibilities, verify that the staff member is working within the framework of an appropriate role load, and reaffirm the staff member's authority to self-schedule assignments that are not essential components of a shared schedule.

Feedback

One of the most fundamental components of sound job design is adequate and appropriately-timed feedback to enable self-assessment of job performance. The employee should have multiple sources of feedback, perhaps including coworkers as well as a supervisor or manager. Feedback is ideally related to but not entirely a component of periodic employee performance evaluations. The ideal feedback system is structured such that the outcome of a performance evaluation is never a surprise for anyone concerned.

It is especially important that the job itself provide feedback the staff member should be able to independently determine how well he or she is doing the job. One of the benefits of reference work is that some forms of feedback are immediate and continuous. The recognition that a question has been answered to the benefit of a patron or the satisfaction of teaching a patron how to do something new carries an instant reward. One of the impacts of the current fast pace of the reference environment, however, is an increasing sense of uncertainty: Did I provide the answer the patron really needed? Did the patron really understand? Did I spend too long with one patron at the expense of another? Should I have answered the telephone rather than approaching a patron who appeared to be frustrated with the computer? These concerns, which are major contributors to technostress, must be addressed through the development of formative management feedback mechanisms that let the staff member know how well he or she is performing.

Social Contact

Although it has been about two decades since the expression "the electronic cottage" was used to describe what is now termed *telecommuting*, there has been no general move toward the projected environment in which employees would work from their homes with no need to reside in any specific locale or to engage in face-to-face interaction with other employees. For many employees in any environment social interaction is an essential component of work. There are many natural and inherently productive situations in which employees need to interact with each other and with the public in ways that actively mix the combination of "doing the job" and social interaction.

People are drawn to public service occupations such as librarianship at least in part by their desire to interact with other human beings. Unfortunately, one of the impacts of advancing technology is a temptation to withdraw from direct human contact in favor of digital contact. Too many reference desks have been redesigned to make it possible for the reference librarian to have maximum access to electronic resources without leaving his or her chair. Just as patrons are subject to technological idolatry, so are reference librarians potentially prone to arranging their work environments such that every electronic resource is within arms' length. Electronic information resources then become a ready excuse for avoiding the active mode of reference service sometimes known as "pointing with your feet". This is an example of a non-technological issue being attributed to technology. The impact is a situation in which the human inter-action that drew the librarian to the profession is replaced by an interactive setting focusing on the computer in which the patron frequently feels that he or she is no more than an unwelcome intrusion. An important set of manage-ment tasks involves ensuring that reference staff have the opportunity to interact with other people and at the same time making sure that reference staff take appropriate advantage of that opportunity.

Identity, Responsibility, Skill, and Achievement

Identity has to do with the extent to which the staff member understands his or her job as a cohesive whole. In order for the staff member to identify with the job, the job must make sense. A job that is perceived as an array of unre-lated and unstructured tasks will usually be unrewarding and far from fulfilling. Identity is closely tied to autonomy – the staff member must have some oppor-tunity to self-define his or her job in order to establish a personal sense of what the job really is and how he or she contributes to and is benefitted by the job.

Advancing technology can either raise or lower the levels of responsibility and skill required of staff members. Deskilling – the process whereby an employee is required to employ less expertise or skill than in the past – is particularly stressful for the employee affected. An increase in the skill level necessary to perform assigned tasks may carry with it the perception that there is an accompanying increase in responsibility, which may not be welcome. The introduction of word processing hardware and software into secretarial and cler-ical jobs, for instance, was sometimes accompanied by active resistance from employees who felt that they would be required to perform advanced tasks for which they were neither prepared nor compensated. The job design challenge is to match appropriate levels of responsibility and skill to staff members' needs and abilities in a changing environment.

Every employee wants to feel that he or she is accomplishing something; that the tasks performed make a difference and contribute something positive. The need for this sense of achievement is one of the motivators for ensuring adequate feedback to the staff member. Although one of the benefits of reference work is continuous feedback that may instill a sense of achievement, that feeling of accomplishment can be undermined by the lack of clear indicators of accumulated achievement. At the end of the day reference librarians frequently feel a sense of "where did the time go?" that is in no way answered by the knowledge that tomorrow will probably be very much the same. Fostering a meaningful sense of achievement requires building a shared understanding of the role of the staff member within the reference environment, the role of reference services within the library, and the goals that the library in general and the reference operation in particular are striving to achieve. This need to link goals to achievement underscores the need to foster staff cooperation and build a joint vision of the library and its purposes.

The annual or periodic staff member-supervisor performance review and discussion is an essential component in helping each staff member develop and maintain a sense of identity; retain a sustainable, understandable, and acceptable feeling of valued responsibility and skill; and build a self-recognized record of achievement. Additionally, reference staff members need to be confident that they can approach their supervisors when they begin to feel reduced comfort levels in any of these areas. A reference staff member who is experiencing an excessive and unmanageable workload or a burgeoning role conflict has a legitimate and immediate demand on his or her supervisor for counseling and possibly intervention. Every staff member should be able to approach a supervisor with no fear of prejudice to say, in effect, "I don't feel that I know what I'm doing and I don't know what to do about it".

Opportunities to Learn and Develop

Every staff member needs access to learning opportunities that will lead to increased skill and foster a sense of achievement. Historically, staff development in libraries was informal, frequently ad hoc, and generally dependent on the initiative of the staff member rather than on planning and support from the library's administration. Frequently, staff members were expected to learn, grow, and develop on their own time and with their own personal support. Fortunately, recent years have seen a dramatic increase in the recognition that staff members cannot be solely responsible for their own continuing education and training. Opportunities to attend workshops, conferences, or seminars are constantly expanding. This is in itself potentially problematic in that selecting the most advantageous experience can be a challenge.

Managers must recognize that staff members have two basic sets of learning needs. The first is the need for structured experiences, typically of a group nature, that serve as efficient approaches to exploring new areas of content and skill or discussing issues and concerns. In-house staff development sessions provide opportunities not only to explore new content areas, but can also serve as a vehicle for interpersonal interaction, sharing of experiences, and recognition and appreciation of varied skills and perspectives. When used effectively, these experiences combine to provide support for affective as well as cognitive development.

The second need is for individualized learning opportunities, frequently of a hands-on nature, that are timed to the needs of individual staff members. Every staff member's continuing development program should include appropriate aspects of both structured group experiences and opportunities for independent learning.

Role Load

Workload refers to the volume of work an individual is required or expected to accomplish, including expectations that are self-imposed. Role load has to do with the number and complexity of tasks that can be addressed and successfully performed; this varies greatly among individual staff members. "Everyone has an optimal rate of working, and it is desirable that people should be allowed to regulate their own work" (Morris & Dyer, 1998, 287–288). Both job performance and job satisfaction are linked to the extent to which the staff member feels in control of the pace of the job. Morris and Dyer discuss both role underload – in which the staff member must work more slowly than the optimal pace – and role overload. The typical situation in a reference setting is role overload. Reference staff members generally report that they have too many tasks to accomplish and too little time in which to accomplish them. One of the real dangers of a highly technologized work environment is the tendency of the volume and rate of demand to drive the pace at which staff members work. The result is that reference staff legitimately complain that they are stretched to the limit, never able to take a break, and perpetually exhausted.

Role Conflict
Role conflicts arise when the individual staff member has difficulty determining priorities among roles; when performing any one task, the staff member feels constant pressure from the other tasks that are not being performed. Role conflict has been discussed in the context of the expanding roles of the reference

librarian, which include question answerer, teacher, document provider, program speaker, manager, and more. Faculty status for college and university librarians is a frequent invitation to the development of role conflicts. The priorities of professional and faculty roles must be clearly defined, associated with predictable rewards, understood by the librarian, and accepted by the librarian. When desk time is perceived as stealing time from faculty responsibilities or when faculty responsibilities are perceived as something to be done on personal time, severe role conflict can arise. Excessive role conflict is a prime contributor to burnout. A management strategy for minimizing role conflict is to require staff members to engage in periodic self-assessment and description of their jobs, to explicitly identify priorities and goals, and to develop personalized strategies and tactics for achieving an appropriate balance of roles.

Role Ambiguity

Role ambiguity is closely tied to autonomy and shared decision making. The most obvious example of role ambiguity is the situation in which a decision must be made and the supervisor or manager who would normally make the decision is unavailable. There may be reference staff members present who possess the knowledge and analytical skill to make the decision. If there is no established procedure for delegating authority, or if the manager tends to question or challenge decisions made in his or her absence, the need to make a decision in the manager's absence tends to freeze the operation of the reference unit. This is accompanied by frustration, stress, and failure to make a decision that needs to be made.

Rest

Perhaps the most damaging aspect of the increased stress and rapid pace of the electronic reference environment is a tendency for reference staff to avoid taking breaks, not because they are unaware of the need for rest, but because they are fearful of failing to meet patron needs. The need for breaks is closely related to the need for task variety. Every worker needs some time when he or she is not working at all and has no immediate assignment. The natural tendency to take a break only when one feels tired is potentially harmful to the individual and to the performance of his or her job. Breaks are an essential means of preventing fatigue and actually have little power to cure fatigue after its onset. There is some suggestion that a large number of relatively short breaks is superior, at least for some individuals, to a small number of longer breaks. Whenever possible, the staff member should have at least some flexibility in determining when to take a break.

Fig. 2. Attitudes Toward Technology.

ATTITUDES TOWARD TECHNOLOGY

Weil and Rosen (1997) described three major *techno-types*. *Eager Adopters* are the 10–15% of the population who truly and instinctively love and embrace technology. *Hesitant "Prove Its"* comprise 50–60% of the population; these individuals "do not think technology is fun and prefer to wait until a new technology is proven before trying it" (Weil & Rosen, 1997, 18). The third group, *Resisters*, is made up of the 30–40% of the population who "want nothing to do with technology, no matter what anyone says or does to convince them that some of it is useful" (Weil & Rosen, 1997, 19).

The real picture, though, is more complex: people do not divide easily into discrete categories, but tend to possess multiple simultaneous attitudes toward technology. Figure 2 (Wallace and Van Fleet, 1997) illustrates the possible intrapersonal relationships involving the overlapping categories of People Who Accept Technology, People Who Like Technology, and People Who Don't Like Technology. Although no scientific study underlies this analysis, it seems safe to assume that each of the areas within the diagram is in fact represented in the general population, including the possibly unexpected categories of People Who Like and Don't Like Technology and People Who Like Technology but Don't Accept Technology. It should also be noted that any individual's place within this analysis is situational – when everything is working correctly and the technology is meeting or exceeding individual needs or expectations, liking and accepting the technology are easy. When the software doesn't perform as the user feels it should, perhaps due to changes between releases of the product, not accepting the technology can rapidly turn to not liking the technology. In addition to providing a graphic illustration of the variable and uncertain nature of attitudes toward technology, Fig. 2 points to one of the most essential tools for use in coping with and managing technostress – a sense of humor.

REFERENCES

Alesandrini, K. (1992). *Survive Information Overload: The 7 Best Ways to Manage Your Workload by Seeing the Big Picture*. Homewood, IL: Business One Irwin.

Armour, S. (1999). Rage Against the Machine: Technology's Burps Give Workers Heartburn. *U.S.A Today*, August 16, B1–B2.

Arnetz, B. B., & Wiholm, C. (1997). Technological Stress: Psychophysiological Symptoms in Modern Offices. *Journal of Psychosomatic Research, 43*, 35–42.

Ashley, W. C., & Morrison, J. L. (1997). Anticipatory Management: Tools for Better Decision Making. *The Futurist, 31*, 47–50.

Baldwin, B. A. (1985). *It's All in Your Head: Lifestyle Management for Busy People*. Wilmington, NC: Direction Dynamics.

Bamberg, E., & Busch, C. (1996). Worksite Health Promotion by Stress Management Trainings – a Metaanalysis of Experimental Studies. *Zeitschrift Fur Arbeits-und Organisations psychologie, 40*, 127–137.

Barbalet, J. M. (1999). Boredom and Social Meaning. *British Journal of Sociology, 50*, 631–646.

Bell, D. (1973). *The Coming of Post-Industrial Society: A Venture in Social Forecasting*. New York: Basic Books.

Bennis, W. G., Benne, K. D., & Chin, R. (1985). *The Planning of Change*. New York: Holt, Rinehart and Winston.

Brod, G. (1984). *Technostress: The Human Cost of the Computer Revolution*. Reading, MA: Addison-Wesley.

Bush, V. (1945). As We May Think. *Atlantic Monthly, 176*, 101–108.

Churbuck, D. C. (1993). Good-bye, Dewey Decimals. *Forbes, 151*, 204–205.

Corsini, R. J. (1999). *The Dictionary of Psychology*. Ann Arbor: Braun-Brumfield.

Everly, G. S. Jr. (1989). *A Clinical Guide to the Treatment of the Human Stress Response*. New York: Plenum.

Flower, J. (1996). *The Change Project*. http://www.well.com/user/bbear/change1.html. Last accessed April 27, 2000.

Freedy, J. R., & Hobfoll, S. E. (1994). Stress Inoculation for Reduction of Burnout A Conservation of Resources Approach. *Anxiety, Stress, and Coping, 6*, 311–325.

Gorkin, M. (1998). *The Four Stages of Burnout*. Mental Health Net. http://mentalhelp.net/articles/stress10.htm. Last accessed April 27, 2000.

Hudiburg, R. A. (1996). Assessing and Managing Technostress. (Talk delivered to the Association of College & Research Libraries Instructional Section, American Library Association Annual Conference, July 8, New York).

Klapp, O. E. (1986). *Overload and Boredom: Essays on the Quality of Life in the Information Society*. New York: Greenwood.

Knowles, M. S. (1980). *The Modern Practice of Adult Education, revised and updates*. Chicago: Association Press.

Koch, R. (1998). *The 80/20 Principle: The Secret of Achieving More With Less*. New York: Doubleday.

Kupersmith, J. (1992). Technostress and the Reference Librarian. *Reference Services Review, 20*(7–14), 50.

Long, B. C., & Flood, K. R. (1993). Coping with Work Stress Psychological Benefits of Exercise. *Work and Stress, 7*, 109–119.

Lowenthal, R. A. (1990). Preliminary Indications of the Relationship between Reference Morale and Performance. *RQ, 29*, 380–393.

McGrath, P. (1999). Potholes on the Road Ahead: Tense About the Future? You're Not Alone. *Newsweek, 134*, 78.

Maslach, C., Jackson, S. E., & Leiter, M. P. (1997). Maslach Burnout Inventory: Third Edition. In: C. P. Zalaquett & R. J. Wood (Eds), *Evaluating Stress: a Book of Resources* (pp. 191–281). Lanham, MD : Scarecrow.

Meichenbaum, D. (1977). *Cognitive-Behavior Modification: An Integrative Approach*. New York: Plenum.

Moch, J. (1998). Validation of the Stress Curve Against the Maslach Burnout Inventory. (paper presented at the 2nd World Congress on Stress, Melbourne, Australia, October 25–30, 1998).

Monat, A., & Lazarus, R. S. (1991). *Stress and Coping*. (3rd ed.). New York: Columbia University Press.

Morris, A., & Dyer, H. (1998). *Human Aspects of Library Automation*. 2nd ed. Aldershot, Hampshire: Gower.

Otlet, P., & La Fontaine, H. (1895). *Creation d'un Répertoire Bibliographique Universel*. Bruxelles: publisher unknown.

Potter, B. (1993). *Am I Burning Out?* http://www.docpotter.com/Beajob_amI_bo.html Last accessed May 12, 2000.

Rose, P. M., Gray, S. A., & Stoklosa, K. (1998). A Focus Group Approach to Assessing Technostress at the Reference Desk. *Reference & User Services Quarterly, 37*, 311–317.

Schon, D. (1971). *Beyond the Stable State*. New York: Norton.

Shrink, C. (1996). *Burnout Inventory*. http://www.queendom.com/burn1_frm.html Last accessed May 12, 2000.

Smith, N. M., Birch, N. E., & Marchant, M. P. (1984). Stress, Distress, and Burnout: A Survey of Public Reference Librarians. *Public Libraries, 23*, 83–85.

Steiner, P. (1993). Cartoon. *The New Yorker, 69*, 61.

TechnoStress Measurement Tools. (n.d.). http://www.technostress.com/WRmeas.html Last accessed May 20, 2000.

Toffler, A. (1970). *Future Shock*. New York: Random House.

Toffler, A. (1990). *Powershift: Knowledge, Wealth, and Violence at the Edge of the 21st Century*. New York: Bantam.

Toffler, A. (1980). *The Third Wave*. New York: Morrow.

Van Fleet, C. (1990). Lifelong Learning Theory and the Provision of Adult Services. In: K. M. Heim & D. P. Wallace (Eds), *Adult Services: An Enduring Focus for Public Libraries* (pp. 166–211). Chicago: American Library Association.

Van Fleet, C. (1999). *The Measuring Library Services Project: Background and Update*. Panel presentation, Measuring and Reporting Electronic Services, with Jay Burton (State Library of Ohio), Steve Wood (Ohio Public Library Information Network), and Chuck Gibson (Worthington Public Library). TechConnections. Columbus, OH. June 1999.

Van Fleet, C., & Wallace, D. P. (1999a). *The Increasing Importance of Digital Information Resources and Their Effect on Reference Service in Public Libraries*. Professional Development Forum, Phoenix (Arizona) Public Library, October 28, (1999). Phoenix, AZ.

Van Fleet, C. and Wallace, D. P. (1999b). *Virtual Libraries – Real Stress: Handling Change at the Reference Desk*. Sponsored by the Maricopa County Library Council and the State Library of Arizona. October 29, 1999. Glendale, AZ.

Wallace, D. P., & Van Fleet, C. (1997). Describing Technological Paradigm Transitions: A Methodological Exploration. *Journal of the American Society for Information Science, 48*, 185.

Weil, M. M., & Rosen, L. D. (1997). *TechnoStress: Coping With Technology @Work @Home @Play*. New York: John Wiley & Sons.

Wells, H. G. (1938). *World Brain*. Garden City, NY: Doubleday, Doran and Company.

APPENDIX

Reference Stress Inventory

Instructions: mark the appropriate box for the extent to which you agree with each of the following statements.

	1 Strongly Agree	2 Agree	3 Neutral	4 Disagree	5 Strongly Disagree
1. I have adequate time to do my job well.					
2. I understand all the tasks necessary to do my job.					
3. I feel close to the patrons I serve.					
4. I feel confident of my ability to do my job.					
5. As the development of technology advances, my work gets easier.					
6. I can easily teach others how to use the computer.					
7. There are enough people working in my department to get the job done.					
8. I never feel that I have to apologize to patrons.					
9. I usually find the right answer.					
10. I'm a good user of electronic information sources.					
11. The Internet has brought me closer to the patrons I serve.					

	1 Strongly Agree	2 Agree	3 Neutral	4 Disagree	5 Strongly Disagree
12. Library service has been improved by computer information resources.					
13. Computer information has enhanced my ability to do a good job.					
14. No one has really been replaced or displaced by computers.					
15. I rarely have a problem with the computer.					
16. The training I have received in new technologies has been excellent.					
17. I know I'm providing good service to patrons.					
18. I have important information and advice to share with patrons.					
19. Using the Internet to find information is an essential part of what I do.					
20. I use Internet resources in my work a lot.					
21. My supervisor is supportive of me and my co-workers.					
22. Everyone should know how to use the Internet.					
23. Helping patrons learn new information resources is an exciting challenge.					

	1 Strongly Agree	2 Agree	3 Neutral	4 Disagree	5 Strongly Disagree
24. In my library, when we have technological difficulties, technical support is excellent.					
25. I have little difficulty determining how to best help patrons.					
26. I rarely make mistakes when using the computer.					
27. Computer resources make my job easier.					
28. Technology allows me more control over my workday.					
29. Libraries should move into new areas of technology as quickly as they can afford to.					
30. Technology makes my life less stressful.					
31. Computers allow me to produce more work.					
32. Computers make my work production more accurate.					
33. Patrons recognize my ability to find the best source to meet their needs.					
34. When I need to learn a new technology I look forward to learning it.					
35. The quality of my library's technology training program is excellent.					
36. The impact of technology on the health of library employees has been positive.					

	1 Strongly Agree	2 Agree	3 Neutral	4 Disagree	5 Strongly Disagree
37. My job provides me with adequate opportunities to learn new things.					
38. I am satisfied with my ability to do my job well.					
39. I enjoy teaching patrons how to search online and CD-ROM databases.					
40. As new technology is added, the speed with which I accomplish work almost always increases.					
41. People are the masters of technology and technology is a tool we are using wisely.					
42. I like the people I work with.					
43. I have time to provide patrons with the help they need.					
44. It's important to help patrons learn how to use the computer.					
45. It's easy to learn new Internet search tools.					
46. Public access to the Internet is an essential library service.					
47. Internet users understand the value of libraries.					
48. I like working with online information sources.					
49. Technology hasn't been responsible for any personnel changes in my library.					

	1 Strongly Agree	2 Agree	3 Neutral	4 Disagree	5 Strongly Disagree
50. My department works well as a team.					
Column Totals	Number of checks in this column =	Number of checks in this column ×2 =	Number of checks in this column ×3 =	Number of checks in this column ×4 =	Number of checks in this column ×5 =
GRAND TOTAL	Sum of scores for all columns =				

Scoring the Reference Stress Inventory

If your Grand Total is 100 or less, you are not a sufferer from reference stress. Pursue preventive maintenance.

If your Grand Total is between 101 and 150, you are a potential casualty of reference stress. Step up your routine preventive maintenance activities and monitor for further signs of stress.

If your Grand Total is 151 or above, you may be suffering from reference stress. Begin a program of assessment and intervention immediately.

If you scored highest on questions 21, 23, 25, 37, 38, 42, and 50, the sources of stress in your job may not be related to technology. Think about ways in which you can work to improve your general job satisfaction.

If you scored highest on other questions, you may be suffering from technostress. Think in terms of specific approaches to reducing technology-induced stress in your working environment.

Reference Stress Inventory © 1999 by Connie Van Fleet and Danny P. Wallace; individual librarians and libraries are granted permission to use the *Reference Stress Inventory* with full and proper attribution.

DEVELOPING A STRATEGIC PLAN FOR INTEGRATED INFORMATION RESOURCES AND SERVICES

Stephan R. Reynolds

PREFACE

In 1998, Nichols College in Dudley, MA, reorganized the reporting structure of its library services (including audio/visual services), computing services (including academic computing, administrative computing, database administration, and network administration), and telecommunication services (including telephony, voice mail, satellite operations, and the campus FM radio station operations). Under the new name of Information Resources and Services (IRS), the campus library, the information technology (IT) department (formerly known as Computer Services), and the telecommunications department were placed under the direct leadership and responsibility of the institution's chief information officer (CIO).

The reorganization was initially mandated by the president of the college without the benefit of any strategic plan designed to ensure the successful integration of those services. Immediately following the reorganization, and at the request of the CIO, a committee-driven project was initiated that culminated in the development of a five-year strategic plan designed for the successful integration of library services, computing services, and telecommunication services at Nichols College.

This article describes much of the research that served to define the strategic plan for integration of these information resources and services. Also included

Advances in Library Administration and Organization, Volume 18, pages 45–121.
Copyright © 2001 by Elsevier Science Ltd.
All rights of reproduction in any form reserved.
ISBN: 0-7623-0718-8

45

in this discussion are political, logistical, and technical issues that impelled considerations for the decision to create a centralization of information resources services at Nichols College. This discussion should serve the reader by supporting the realization that there is no single "right" answer to the question of whether or not an academic institution should encourage and implement such a reorganization. Clearly, the correct organizational model for any academic institution is one that effectively provides information resources and services upon demand to all campus constituents.

OVERVIEW

During the decade preceding this project, lines between information content and information delivery had become less distinct and had begun to merge at Nichols College. Although the strengths of those individuals supporting library services, computer services, and telecommunications services were extensive, the level of service effectiveness needed to be improved. The problem was that the level of effectiveness in providing information resources and services did not match the level of effectiveness projected in the college's vision.

The purpose of this project was to develop a strategic plan for the integration of library services, computing services, and telecommunication services. The first research question for this study was "What are the practical benefits of organizational integration?" Second, "What cultural concerns and differences affect organizational integration?" Third, "What internal processes require reengineering to promote successful restructuring?" Fourth, "What are the specific steps for a successful organizational integration?"

The development problem-solving methodology was utilized for this project. Six procedures were used to conduct this project. A review was conducted of literature that related to the integration of these services at similar academic institutions, the demand for technology in college libraries, and the leadership of successful organizational change. A formative committee discussed the issues, concerns, and practical requirements of integration. The formative committee reviewed the expert literature, defined practical benefits of organizational integration, sought to understand how the effective delivery of information resources could enhance the skills and competencies expected from an academic curriculum, identified internal processes requiring reengineering, established criteria for the strategic plan, and developed an initial draft.

The draft was presented to a critique committee for review and comment. The strategic plan draft was then presented to a summative committee for validation and approval. Based upon the comments of all committees, a final strategic plan was then submitted to the office of the president of the college for consideration, approval, and immediate implementation.

INTRODUCTION

Nichols College completed the implementation of a high-speed fiber-optic voice and data infrastructure in 1997. This network was designed to support its newly acquired telecommunications system, voice mail, Internet access, electronic mail, and data environments well into the future. With this infrastructure available to all students, faculty, and staff, the ability for users to access information upon demand was greatly enhanced.

Also in 1997, the Computer Services department at Nichols College was renamed Information Technology (IT). The rationale for this change was based upon the belief that technology had become so pervasive and so diverse that the term "computer services" had become passé. Over the years, the IT department had become responsible for the complete workings and the seamless integration of communications, computer systems, and the network infrastructure. IT had also been responsible for the service and support required to back up each of those information endeavors.

Furthermore, IT had served Nichols College by providing for the security, retention, backup, and recovery of electronic information and voice data. IT had also maintained automated information systems and provided training, technical consultation, and planning in the application of information technology in both the academic computing and administrative computing environments.

The library at Nichols College had been an important part of the history of the college. Since its beginning, the primary mission of the library had been to support and enrich the academic programs and teaching curriculums of Nichols College. The library offered a diverse array of materials for the intellectual, professional, and cultural development of the students, faculty, and staff. These materials, whether printed or electronic, addressed the information needs of the entire Nichols College learning community.

Nature of the Problem

Up to this point, Nichols College had taken a traditional approach to the organization's functional structure; the campus phone system was the responsibility of the chief financial officer (CFO), all library services were the responsibility of the chief academic officer (CAO), and computer-related support was the responsibility of the director of computer services, who reported to the CFO. In 1997, the position of chief information officer (CIO) was created at Nichols

College. A national search was conducted, and within a few months, the college's first CIO was hired.

The CIO's initial responsibilities included all computer-related support for both academic and administrative computing environments. Campus telecommunications, including both telephony and voice mail, became a part of the CIO's responsibilities. Telecommunication services had formerly been the responsibility of the chief financial officer. The campus cable television channels as well as all multimedia services, formerly the responsibility of the library director, became the responsibility of the CIO. Also, the campus FM radio station operations, formerly the responsibility of the dean of student affairs, became a part of the CIO's responsibilities. However, all library services, including electronic information resources, were left under the leadership and responsibility of the chief academic officer.

Since the beginning of this decade, a growing number of colleges and universities have developed and implemented strategic plans for the integration of library services and computer services for a variety of reasons. Arising from the growth of electronic information resources and the pervasiveness of the technology now required to deliver those resources, many academic institutions are looking at the political, logistical, and technical benefits of this type of integration as a part of their strategic long-term vision and planning.

During the 1997–98 academic year, many positive changes occurred at Nichols College. When the president of Nichols College chose not to renew his contract, the board of trustees unanimously elected the academic dean at Nichols College as its first female and sixth president. Soon after, the new president charged the CIO with examining better ways to provide information upon demand to the campus community. The president believed that the college needed to fully-integrate its information resources and services group, and combine the strengths of those individuals supporting library services, computer services, and telecommunications services under a single leader to provide a unified vision as to how best to serve the information and computing needs of the students, faculty, and staff. Although those individuals supporting library services, computer services, and telecommunications services were strong, their combined level of service effectiveness needed improvement. The problem was that the college's effectiveness in providing information resources and services did not match the level of effectiveness projected in the college's vision.

Purpose of the Project

The purpose of this project was to develop a strategic plan for the integration of library services, computing services, and telecommunication services at

Nichols College. This strategic plan would then be presented to the college's president as a recommended method for increasing the effectiveness of information delivery to the campus community.

Background and Significance of the Problem

During the 1980s, Nichols College initiated the requirement that all students own and utilize laptop computers to enhance the learning environment. Since that time, Nichols College had made a multi-million dollar investment in a complex, state-of-the-art technological infrastructure accessible to students, faculty, and staff to support lifetime learning. The institution continued to develop support structures within both the library and computing organizations to help users effectively utilize the information resources that continually became available. From then on the principle that information resources and services must be fundamentally integrated and aligned with the mission of Nichols College ruled all thinking, planning, and curriculum development.

The development of a strategic plan for the integration of library services, computer services, and telecommunications services occurred within the intent and spirit of that mission. This organizational change was designed to increase the level of effectiveness in providing information resources and services to all campus constituents. Without this strategic plan, it was believed that, given the budgetary resources available, the library would continue to serve students by doing little more than acquiring and archiving information. However, through the successful integration of information resources and services, the library could evolve from being a place where information was kept to become a portal through which learning students and professionals could access the vast information resources available in today's world.

Student and faculty expectations were such that there was a growing demand for more sophisticated information resources and services, which required much broader access to data. In responding to this pressure for greater information access, the organization needed to adopt a more consumer or customer orientation. A strategic plan for the integration of these services was developed to help codify successful processes and practices intrinsic to the institution and blend them with the best customer-oriented organizational models.

It was believed that this convergence of information and the technology upon which it relies would help students master a body of knowledge and develop greater critical thinking skills. This kind of information literacy program could assist the learner in locating, extracting, and processing information in order to generate a desired knowledge base. The general belief at Nichols College was that colleges and universities have a fundamental responsibility to develop strategic plans that can establish a solid foundation for the creation, implementation, utilization, and delivery of educational opportunities. The thought and effort that went into the development of this strategic plan would help to define the success of information resources and services at Nichols College during the next decade and beyond.

Research Questions

There were four research questions posed in this project. First, "What are the practical benefits of organizational integration?" Second, "What cultural concerns and differences affect organizational integration?" Third, "What internal processes require reengineering to promote successful restructuring?" Fourth, "What are the specific steps for a successful organizational integration?"

REVIEW OF LITERATURE

A review of expert literature on the subject of the reorganization and integration of college and university computing services and library services was conducted in three major topic areas. First, current models of organizational integration implemented at similar institutions were studied, and the cultural differences and similarities between computer services and library services were identified. Next, the demand for technology in academic libraries as a means for achieving the information literacy and information competency of students was analyzed. Finally, the leadership of successful organizational change within academia was reviewed.

Introduction

During the decade of the 90s, a growing number of colleges have developed strategic plans for the integration of computer and library services. Arising from the growth of information resources, the pervasiveness of the technology now required to deliver those resources, and the need to elevate a college's present level of effectiveness in providing information resources and services, many academic institutions view integration as an element of their strategic long-term vision. Hardesty (1999) believes that integration can strengthen both service-oriented units. In describing a typical integrated structure, he writes:

> The individual in charge of both [units] usually holds a senior administrative appointment, reporting directly to the president and serving on the senior administrative council. This position can enhance visibility, attract resources, and bring added benefits to both units and their staffs. In some schools, the two units work so well together that the potential for cooperation and service enhancement can be realized only with integration (p. 1).

Nichols College recognized that the current concentration within the Information Technology department on reengineering various campus activities (with an emphasis on the retrieval of relevant data for various constituents) was congruent with the library's client-centered and information-related mission. Furthermore, it was believed that this synergy should be used to increase the level of effectiveness in providing information resources and services to the students, faculty, and staff who work and study within the Nichols College learning community.

However, the successful integration of computing services and library services at colleges and universities does not simply require the merger of the two functions, but rather creates an operational environment where the strengths of its individual members complement each other and are focused on accomplishing a common institutional mission. Hirshon (1998), a pioneer in the collaborative integration of computing services and library services, states:

> Integrating operations is a tool to achieving institutional objectives, not an objective in itself. The institution must have a vision of what an integrated organization will accomplish, and how the integration will help to achieve that vision or reach the established goals (p. v).

Beneath the rhetoric and experimentation, there is a growing recognition that the traditional boundaries of information and technology are becoming less distinct. Young (1994), the Director of University Libraries at Northern Illinois University, believes that the concentration within the information technology area on reengineering various campus activities, with an emphasis on the retrieval of relevant data for various constituents, is congruent with the library's client-centered and information-related mission. He observed that the mission relatedness of academic libraries and computing services, together with the desired goal of coordinated and effective access to information resources, transcends matters of control and culture.

Current Models of Organizational Integration

Many colleges and universities have begun to address issues relating to their effectiveness in providing information resources and services. Some are integrating library services and computer services to ensure greater collaboration between information resource and service providers within the organization. Hirshon (1998, pp. 35–37) compiled a list of institutions with integrated computing and library operations, and much was learned from this new field of experts.

Beginning in 1994, Gettysburg College, a small liberal arts college located in Gettysburg, Pennsylvania, merged the Computing Services Department and the Library into a single division called Strategic Information Resources.

Aebersold and Haaland (1994) contend that, prior to the merger, Gettysburg College had been devoting significant resources to information technology (as it was known at the time) in such functional areas as the library, computer services, telecommunications, management information systems, and the print shop. These were all areas responsible for helping members of the college community gain access to, store, retrieve, and analyze information. Aebersold and Haaland assert that, because a well-founded liberal arts curriculum has proven effective in helping students master new technologies and synthesize new material, the merger advanced the effective development of strategies designed to promote the use of advanced information technologies in the college's instructional programs.

Lehigh University, located in Bethlehem, Pennsylvania, began a reorganization of its information services in 1997. The merger was an effort to succinctly examine more effective methods of providing information resources and services to their customers. Foley writes "Lehigh University Information Resources must constantly seek strategic opportunities to improve client services, employ the most effective technology for the task to be accomplished, and keep the costs for the University at a minimum" (Foley, 1997: p. 2).

Hirshon (1996), who along with Foley helped bring about the restructuring of information resources and services at Lehigh University, believes that a well-founded strategy for integration is one that encompasses all client information technology needs and explores how to meet those needs with both traditional and emerging technologies. According to Hirshon, the integration of information services is essential in helping to understand the full costs and potential benefits of delivering the information access demanded by users.

The University of Arizona, in a strategic plan that describes a 21st century electronic environment that reaches its constituents at any time in any place, identified information technology as an indispensable and strategic part of the university's direction. Detloff, Ecelbarger, and Wilburn (1996), administrators at the University of Arizona, state:

> Our vision requires integrating data from many different sources, expanding its accessibility, and building a secure technical foundation that is usable, reliable, extensible, portable, adaptable, and manageable. We selected an architecture that allows us to capitalize on our systems and expertise in the transition to an integrated information systems environment (p. 1).

Cisneros, Hunt, and McCollam (1997), system technologists at the University of Arizona, agree with the strategy of restructuring the campus information environment. They believe that information repositories containing redundant data will result when there is no overall integrated information architecture in place to orchestrate the acquisition, organization, automation, and preservation of information.

Similarly, Virginia Polytechnic Institute and State University (Virginia Tech), a research university located in Blacksburg, Virginia, has recently harnessed the collaborative spirit of an integrated library and information systems team, and has successfully launched a program that provides electronic library resources for extended campus learners. Eustis and McMillan (1997), key members of the team developing this approach, contend that there is an opportunity for colleges and universities to restructure themselves through the delivery and use of technology into vital organizations that will enhance student learning and meet the challenges of the next century.

In 1995, the Massachusetts Institute of Technology initiated a process to reengineer many of its information resources and services functions. Goguen and Hogue (1997), in describing this project, conclude that the key elements of any reorganization are the information processes involved. They believe that these processes include discovery, delivery, integration, service, and support.

In 1994, Brooklyn College merged academic computing and the library under a single directorate. Writing about the merger, Higginbotham (1997) explains that the process of integrating these services allowed the college to reexamine the structure and processes supporting the delivery of computing services on campus. In doing so, Brooklyn College was able to have a positive impact upon its effectiveness in providing information resources and services to the campus constituency.

Given the campus-wide understanding that information is a resource required by all, Hartwick College, a small liberal arts institution located in Oneonta, New York, began a concerted effort in 1996 to integrate information resources by providing pervasive tools and access to the campus community. In a report describing this initiative, Conley, Detweiler, Falduto, and Golden (1996) assert that it is the sharing, comparing, and mutual supporting that make information integration succeed in higher education.

The University of North Carolina at Charlotte recently completed the implementation phase of a combined user support operation for computing and library services. Mancuso and Stahl (1996), part of the implementation team, maintain that the integration of information resources and services has

> ... made concrete to the staff and the campus in many ways what we mean when we say that this project creates a "user-centric" organization. It has caused us to reconsider many of our beliefs and organizational models. Most importantly, though, we believe that it is what we need to do to support the university in the most effective way that we can (p. 8).

The strategic five-year plan at Grambling State University, Grambling, Louisiana, describes a phased integration of services pursuant to the need to become more responsive to the institution's information needs. Beginning with a plan to link and integrate its disjointed information resource systems, Carter,

Lundy, and Penn (1993) agree that the attainment of virtually all of its institutional goals is dependent on an improved technological base. Through the integration of resources and services, the college is embracing the philosophy of user-driven and distributed computing, the decentralization of access to information, and the establishment of a seamless information resource environment.

Duwe, Luker, and Pinkerton (1995) describe a similar strategy implemented at the University of Wisconsin at Madison. There, technology-related units were integrated into a single division. The top three issues were to improve information technology services to students, facilitate access to information resources, and establish an integrated technology architecture.

DeSantis and Laudato (1995) describe the University of Pittsburgh's unique approach to designing an enterprise-wide, integrated information architecture that tightly couples the information architecture with efforts to reengineer the workplace. As state and local governments strive to make gains in the quality of information content and efficiency of information delivery, the State University of New York has committed to seeking methods of integration of its information resources and services. Blunt (1993) writes:

> ... to lead in this field, we will pursue creative new ways of applying technologies directly to practical problems of information management and service delivery, focusing on increasing productivity, reducing costs, increasing coordination, and enhancing the quality of operations and services (p. 3).

Collaboration between computing and library services staff is the ultimate goal of integrating resources. Bernbom and Hess (1994) believe that between these functional groups, areas of parallel interest exist where technologists and librarians are working on the same, similar, or analogous problems of information technology or information management. Pflueger (1995), the chief information officer at California Lutheran University, asserts that, as the university continues to define and refine its vision for infusing sophisticated, user-centered technology into all aspects of university life, the greatest inducement for collaboration between computing and library services is the university's commitment to technology and its desire to get the most out of its investment.

As users of information demanded more access to the computer-based data, faster response on system problems, and development of new information systems, Glenn and Mechley (1993) write that a strategy involving the integration of information resources and services developed at Cincinnati Technical College created an environment that empowered users through the collaborative ingredients of computer literacy, integrated information systems, and information resources query capability. The integration of services resulted in increased user satisfaction.

Conrad, Rome and Wasileski (1993), administrators at Arizona State University, contend that the decentralization of traditional library and computing units through integration allows users greater access to support skills spread across a collaborative information resources and services group. Eaton and Schuler (1994), system administrators at the University of Saskatchewan, support this idea as well, stating that decentralization through integration is necessary in order to accommodate the growing ability of information consumers to process data themselves. Saskatchewan's organizational model addresses the needs of an integrated resources and services group designed to create shared information services rather than shared information systems.

George Mason University has built a new technology infrastructure that serves as a catalyst for changing the teaching and learning environment. Gabel, Hurt, O'Connor and Sevon (1995) describe this new integrated computing and library services environment as deriving from a desire to create a "teaching library" to help students achieve information literacy, and to use technology to achieve academic excellence as an alternative to relying on the development of a traditional paper library collection. Penrod (1996) documents similar reasons for integration of services at the University of Memphis.

The University of Kentucky library is undergoing a change from a traditional hierarchical organizational structure to a flattened team-based organization. According to Molinaro (1997), the impetus for this change revolves around the desire to become a more user-centered organization supported by new technological opportunities. Beltrametti (1993) describes an integrated information resources and services organization that enables the University of Alberta community to share computing resources distributed across campus. Likewise, Arzt, Beck, Gordon and May (1993) chronicle the integration of computing and library resources at the University of Pennsylvania.

Clearly, colleges and universities across the country are pushing to achieve a higher level of effectiveness in providing information resources and services. Many of these institutions have adopted a strategy of integrating of computing services and library services in order to accomplish their goal of effective user support. Based upon her experiences, Sharrow (1995), University Librarian at the University of California at Davis, where an integration of information resources and services has taken place, concludes:

> Both librarians and information technologists want to deliver the appropriate information in the appropriate ways and at the appropriate time. By working together and avoiding pitfalls, our functional units – and our customers – have everything to gain. We have achieved a full, productive, and enthusiastic partnership between the library and IT. While this achievement is personally satisfying to the staffs of both organizations, its true value obviously lies in the services we are now able to provide to our users (p. 2).

As the boundary between information technology and information content becomes less distinct, Metz (1999), Instructional Librarian and Interim Director of Administrating Computing Services at Carleton College in Northfield, Minnesota, contends that staff members' perceptions of one another influence the success of integration, and these perceptions become basic tenets for underlying cultural differences and similarities between computer services and library services personnel.

Following the beliefs of traditional psychologists, Metz identifies at least three factors that impede collaboration among groups: (a) social distinctions, (b) salary differentiation, and (c) subcultural differences. To specifically show how these factors are relevant in academia, Metz (1999) writes:

> Computing staff come from a variety of educational and experiential backgrounds; they do not share a common professional preparation. In contrast, librarians are acculturated into a common set of values during their professional preparation. Library and computing center personnel tend to have different salary and responsibility gradations. Benefit packages may differ to include, for example, faculty status for librarians but not for computing staff. Finally, more than two cultures exist. Among campus computing staff, academic and administrative employees may have different outlooks, as may hardware and software support specialists. User services staff in computing organizations have different cultures than programmers or technicians. Librarians, too, have a long history of cultural differences between public services and technical services staff, as well as between librarians and support staff (p. 1).

Kiesler (1994), a professor of social and decision sciences at Carnegie Mellon University, believes that collaborations are more effective when they have a synergistic effect – when the whole is more than the sum of its parts. She contends that in a successful integration, staff can amplify each other's ideas and jointly develop a better plan than the plan of any individual. Hence, Kiesler states that it is a good idea to create an electronic or "virtual" group when, because of distance or organizational or social differences, one would otherwise experience no collaboration at all.

In a summary of survey results from computing and library staff, Metz (1999: 1–2) reports that widely varying perceptions are held by each group (see Fig. 1). It is these perceptions that can influence, for better or worse, how effectively these groups can collaborate. Metz contends that these organizational perceptions – whether right or wrong – often define each culture's understanding of the other.

Furthermore, Metz (1999, pp. 2-3) summarizes that when asked, each group identified what they believed to be unique about their roles in academia (see Fig. 2). Interestingly, for all their self-perceptions of total uniqueness and individuality, substantial similarities were reported.

In a final summarization of survey results, Metz (1999, p. 3) reveals that cultural similarities exist between the two groups. Despite the groups' perception

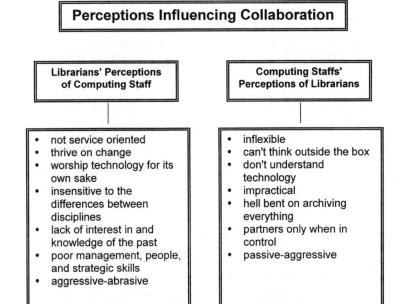

Fig. 1. Adapted from Metz, T. (1999). *Academic libraries and computing centers: The case for collaboration* [Online]. Available: http://library.willamette.edu/home/publications/movtyp/spring99/Metz.html [1999, January 6].

of themselves and of each other, key activities and fundamental problems common to both seem to emerge (see Fig. 3).

Franklin (1994), a director of advanced scientific computing at the University of California, suggests that computing and library staff must capitalize upon their differing perceptions of each other in order to promote collaboration. This collaboration should be based on shared creation and discovery and on mutually agreed upon goals. He contends that:

> Whatever differences of perspective we bring to our joint endeavors are just that: differences of perspective. Our differences are what enrich our participation together. We may be different as individuals or in our organizational responsibilities, but if we understand the setting in which we operate, the commonality of our values and goals becomes much more apparent, as do the collaborative possibilities (p. 23).

According to Creth (1994), a librarian at the University of Iowa, library and computing professionals working together have a unique opportunity to assist

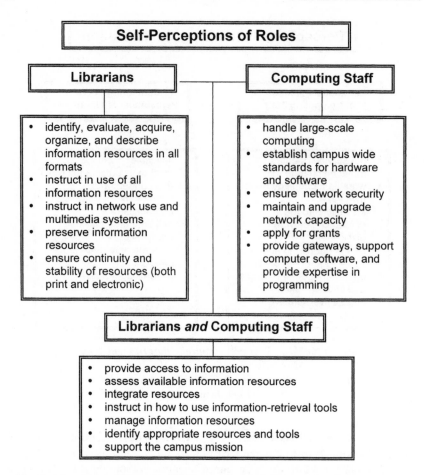

Fig. 2. Adapted from Metz, T. (1999). *Academic libraries and computing centers: The case for collaboration* [Online]. Available: http://library.willamette.edu/ home/publications/movtyp/spring99/Metz.html [1999, January 6].

administrators and faculty. By using information technology, these groups not only create efficiencies with administrative processes but also offer entirely new ways to organize and distribute information and to enhance education.

Demand for Technology in Academic Libraries

Perhaps the greatest factor responsible for the synergy currently seen between computing services and library services is the demand for technology in the

Common Activities and Problems

Activities

- developing training tools and systems documentation
- designing, operating, and using local and wide area networks
- planning, selecting, and operating systems hardware and software
- collecting and organizing information in various forms and formats
- creating, maintaining, querying, and managing databases
- providing consulting and technical assistance
- instructing faculty, students, and staff in all of the above

Problems

- meeting rising user expectations and demands
- understanding new technologies
- revamping services and service procedures
- retraining staff and expanding skill sets
- coping simultaneously with change and the convergence of their responsibilities

Fig. 3. Adapted from Metz, T. (1999). *Academic libraries and computing centers: The case for collaboration* [Online]. Available: http://library.willamette.edu/ home/publications/movtyp/spring99/Metz.html [1999, January 6].

academic library. With the ever-growing electronic availability of information on both internal and external networks, many libraries have turned their focus to providing access on demand rather than building collections. In seeking to enable customers to locate information, librarians are stressing the need to provide resources and services to campus constituents upon demand using all forms of technology.

In today's networked information environment, any library action must be part of a broader campus infrastructure committed to furthering new educational approaches. Rapple (1997), a librarian at Boston College where computing and library services have integrated, suggests:

Above all there must be strong communication and an effective partnership between the institution's library and its computing service. When the two entities remain separate and distinct, the result is often a duplication of effort and a waste of resources. This is poor management of resources and budget, and is also grossly inefficient. Both services need the other in order to attain the same ends for their institution. Librarians need technologists' systems, computing, network, and other technical expertise, while information technologists can learn much from the library's knowledge of users needs (p. 2).

One new demand patrons are placing upon academic libraries is that of information access that transcends space and time. Brentrup, Fark, Jargo and LaMarca (1997), library administrators from Dartmouth College and Brown University, write that the availability and use of electronic information resources change the interactions between patrons and librarians. Patrons no longer have to come to a library or wait for it to be open in order to use an electronic resource. A patron now has the expectation of accessing information anytime from any computer.

Another recent demand for technology in academic libraries relates to technology skills. Librarians are frequently called upon to troubleshoot the existing technology, rather than help a user find suitable content. Stahl (1995) contends that this demand is a self-fulfilling and self-serving effect of technology. According to Stahl, if a customer is required to go to a librarian to find something, then review of that activity needs to be a high priority and ways must be found to remove the requirement. Patrons often view interaction with an intermediary for the purpose of content acquisition as a barrier, while interaction with an intermediary for the purpose of computer support being a value-added service.

The altered and evolving interactions between academic library staff and their patrons have created another demand for technology. As customers are accessing more and more electronic bibliographic and full-text databases, as well as utilizing the vast resources of the Internet from outside the library, librarians need to reach out to them in nontraditional ways to offer the help they need. Massy and Zemsky (1995), administrators from Stanford University and the University of Pennsylvania respectively, refer to this as a basic library tenet of delivering traditional academic value to the customer. As surely as technology changes, so must the process for providing library service to the users of information resources.

Another technological demand on academic libraries involves the World Wide Web's role in academic and administrative operations. In a key strategic planning document prepared for Ball State University, Nasseh (1998) asserts that this new demand embraces the objective of providing opportunities for services, teaching, learning, and research without limitation of time, place, and resources. Nasseh concludes that every higher education institution should realize that web-based information services constitute a vital strategic business investment.

As technology becomes more complex, the infrastructure committed to academic libraries and required to support that technology also becomes more complex. As a director for academic technology at the University of North Carolina, Roberts (1996) contends that the campus network should be viewed by institutions as a strategic asset. The network becomes the path to the learning productivity of the campus community.

The electronic distribution of information is a natural evolution of the use of this infrastructure, and from there the activity of creating or promulgating electronic information quickly follows. Academic libraries have now become the distributor of information online. According to John (1996), user demand now dictates that libraries not only bring more information to their users, but that the information is more up-to-date than before. Being "online" and "on the network" establishes a user expectation that the most current information is accessible.

This demand for technology in academic libraries does not come without a price. Traditionally, academic libraries have followed an approach of local ownership by acquiring substantial serial holdings to meet the needs of their clients. As a result they have always been the largest purchasers of serials and, as such, are tied to the future of large serial publishers. But Brodie and McLean (1995) note that the reduction of operational budgets coupled with the demand for technology is having a significant impact upon this buyer-seller relationship:

> The spiraling price of serials causes libraries to cancel more, which in turn causes publishers to raise the price to cover smaller production runs. At the same time, communications technology and photocopying make it easier to obtain copies of individual articles rather than rely on a serial subscription. This puts further pressure on the publishing industry and means that traditional sources of supply are no longer clear. The changing economics of providing access to the serial literature mean that [academic] libraries can no longer be self-sufficient and are faced with increasing uncertainty in choices for [written] document supply (p. 2).

Clayton (1997), a library administrator at Knox College, Galesburg, Illinois, concedes that in this technology-driven era of distributed computing and access vs. ownership in academia, the end-user's role in determining the content (and subsequently the cost) of information purchased has increased dramatically in recent years. Historically, collections were in buildings, and selection of content was performed by subject librarians and faculty experts. The trend in academic libraries is fast producing an environment where more end-users make the decisions as to what specific information they will be getting for the money they pay for "access" rather than "education".

Another demand caused by the growth of technology in academic libraries is the need for collaboration in the educational process. Never have faculty and students come to rely so heavily upon this interaction. Lowry (1993), a library professional at the University of Iowa, suggested that a blueprint for this collaboration is essential:

> The library is responsible for helping faculty and students to use these resources and to integrate traditional and electronic resources into their research and teaching, for assisting faculty in the development of curricula that incorporate electronic materials, and for assisting students in developing class projects (p. 3).

Furthermore, users of information are demanding that information must be accessible in a "user friendly" manner beyond the constraints of time, place, and content. This, then, brings to bear another technological demand on academic libraries. Aken and Molinaro (1995), library administrators at the University of Kentucky, refer to this as "enhanced access". Since user needs and abilities that comprise information literacy are different, the onus is on providers to present the information in the most unassisted and straightforward manner possible. Providing this good service to users is becoming increasingly difficult as software applications become more diverse and more powerful, and while the expectations of the users are constantly changing.

Technology is, without question, placing increased demands on academic libraries. In many cases, these demands have never before been seen by information professionals. Ernst, Katz, and Sack (1994) reflect on this new reality:

> Three forces of change – organizational, technological, and economic – are underway and gaining momentum in higher education today. Each is prompting discussion, study, frustration, and, in some cases, fear. Taken together, these change forces will alter the nature of higher education. They have brought us face to face with hard choices about how to harness and direct these alterations without becoming their victims (p. 1.).

Historically, academic libraries have been partners in the educational process. Today, that partnership survives, but on a technology-based continuum. Fox (1998), an administrator at the University of Tennessee, suggests that technology will continue to drive this partnership into the future:

> Technology [today] is creating a new educational platform and is reconfiguring the way a student learns. Network learning – accessing libraries, scholars, networks, and information worldwide – is evolving. Education is a discovery process, an exploratory process that provides the widest repertoire of possibilities with which a student is faced when entering a learning situation. Technology in academic libraries can provide this learning expanse, and because of it a student's educational experience will be immeasurably richer (p. 3).

Beyond the mechanics of specific technologies and services lies the desired outcome of information literacy. The American Library Association (1989) has provided a fundamental definition of an information literate person that is accepted by colleges and universities nationwide:

> To be information literate, a person must be able to recognize when information is needed and have the ability to locate, evaluate, and use effectively the needed information. Ultimately, information literate people are those who have learned how to learn. They know how to learn because they know how knowledge is organized, how to find information, and how to use information in such a way that others can learn from them. They are people prepared for lifelong learning, because they can always find the information needed for any task or decision at hand (p. 1).

The development of information literacy closely parallels developments in educational theory. For years, the widely accepted instruction model presented the teacher as the learning authority and the student as the passive learner. Since the 1970s, constructivism has emerged as an alternative instruction and learning model. According to Mayer (1992), constructivist theory emphasizes comprehension and the effective construction of meaning. The learner is not a passive recipient of information, but rather an active participant in the construction of knowledge.

The challenge for colleges and universities then becomes how to create an environment for students that will promote the attainment of information literacy. Teaching facts to students is an inadequate substitute for teaching students how to learn. Students require the skills to be able to locate, evaluate, and effectively use information for any given need. The American Library Association (1989) has challenged college libraries:

> What is called for is not a new information studies curriculum but, rather, a restructuring of the learning process. Textbooks, workbooks, and lectures must yield to a learning process based on the information resources available for learning and problem solving throughout people's lifetimes — to learning experiences that build a lifelong habit of library use. Such a restructuring of the learning process will not only enhance the critical thinking skills of students but will also empower them for lifelong learning and the effective performance of professional and civic responsibilities (p. 7).

Teachers have always assumed the responsibility for teaching literacy to young students in the form of reading and writing skills. Likewise, faculty must share in the responsibility for teaching information literacy. Sonntag (1997) suggests that faculty should be encouraged by librarians to require students to do more than just use the books placed on reserve. It is the role of the librarian to teach the concept that having information is power for life.

It is important that academic libraries foster an environment that nurtures participatory learning in teaching information competencies. Morgan (1998) writes that a viable framework for such an environment is comparable to the rungs of a ladder. The first rung represents data and facts; the second rung represents information; the third rung represents knowledge; the fourth and top rung represents wisdom, or knowledge of a timeless nature. Libraries should enable students to freely move up and down this ladder.

Silverman and Silverman (1999, p. 57) report that the Electronic Enterprise Zone (EEZ), a consortium created between the Technology Based Learning Systems Department of the New York Institute of Technology and various other education and content providers, has developed a paradigm that includes libraries and technology as seamless supports to the learning process. This new paradigm promotes the notion that educators can provide not only resources, but the support and encouragement needed for students to attain high levels of achievement (see Fig. 4). Teachers and community partners collaboratively develop effective and rich learning activities. At core of this paradigm are technology and library systems.

Adding to the demands that technology is bringing to bear on academic libraries is the need for libraries to provide instruction in the area of information literacy. Anderson and Gesin (1997), information administrators at Maricopa Community College in Tempe, Arizona, conclude that library staff are the most appropriate professionals to lead people in finding answers to the challenges presented by the digital environment. In the academic library, teaching information literacy means creating courses and formal instruction with the purpose of encouraging independent learners and critical thinkers.

The success of technology in libraries is paramount to integrating information literacy into the academic curriculum. Lindauer, Oberman, and Wilson (1998, p. 350) present a compilation of core competencies/outcomes necessary to move from information literacy to information competency (see Fig. 5).

Leadership of Successful Organizational Change

Because of the pervasiveness of technology in higher education today, academic institutions are finding that they should not manage through traditional structures and hierarchies but rather through processes created from an information-rich infrastructure. Ernst, Katz and Sack (1994, p. 11) contend that successful organizational change fueled by emergent technologies can be anticipated if the institution: (a) eliminates the technical, cultural, hierarchical, and procedural boundaries that divide or isolated intelligent and motivated people; (b) creates a policy environment that stimulates and rewards collaboration; (c) promotes easy access to the kind of information people need to make sound decisions; and (d) specifies, measures, and rewards the achievement of defined and customer-centric objectives.

Hirshon (1996), who along with Foley helped bring about the restructuring at Lehigh University, believes that the integration of computing and library services will be successful only if there is a shared vision that the change is mutually beneficial. He believes that to be fully effective, restructuring must

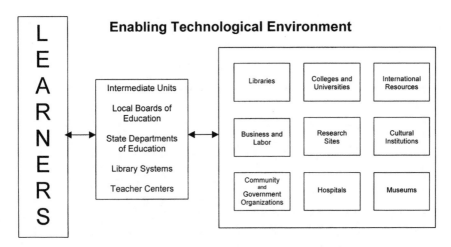

Fig. 4. Adapted from Silverman, S., & Silverman, G. (1999). The Educational Enterprise Zone: Where knowledge comes from! *Technological Horizons in Education Journal, 26*(7), 56–57.

begin with strategic planning, rely upon effective implementation, and must be followed by a continual focus on improving client satisfaction.

Brown (1999), the editor in chief of California Computer News, suggests that this integration of technology has meant fundamental changes to the librarians' traditional roles as providers or facilitators of information. She quotes Mark Parker, a coordinator at the California State Library, as saying "librarians are seeing themselves more in the role of instructor or tutor, teaching people how to use the technology [and] how to use the resources" (p. 41).

Bass (1992) writes that although service should be the main goal of libraries, the goal can often be in conflict with a traditional sense of elitism and self-serving preservation. Consequently, institutions charged with increasing service and quality are being forced to set aside traditional beliefs of organizational leadership. Hawkins (1996), an administrator at Brown University, writes:

> A primary facet of the leadership role is helping staff realize that what is important is service to a mission, not a master. This focus will help all staff to persevere even when they are criticized by demanding faculty, students, and other members of the community (p. 7).

West (1996), an administrator at the University of California, argues that the successful leader is one who assumes responsibility for all aspects of information resources and technology and integrates them into the very fabric of the

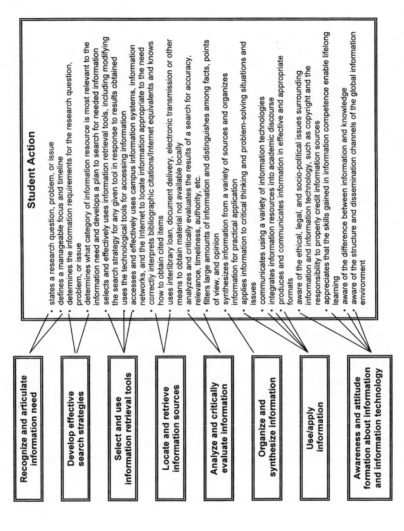

Fig. 5. Adapted from Lindauer, B. G., Oberman, C., & Wilson, B. (1998). Integrating information literacy into the curriculum: How is your library measuring up? *College & Research Libraries News, 59*(5), 347–352.

institution. In an environment where the customer population and demand for services are growing at a steady pace, and where the expectations of those customers are accelerating even faster, Gleason (1996) suggests that there are special challenges to effective communication with the expanding user community. Staff are being asked to address the demands of better service, higher productivity, and lower costs, while at the same time dealing with the pressures of constant change. To balance these seemingly opposite dimensions, Gleason believes that:

> The ideal leader must be an individual who has a good grasp of information technology directions, an understanding of critical issues facing the institution, the ability to solicit and inspire innovation, the skill to evaluate ideas in the context of the big picture, and the instinct to judge when it is appropriate to recommend a change in approach or direction. The ideal leader must be the change agent (p. 2).

LeDuc (1996a), a technology leader at Miami-Dade Community College, argues that today's leaders need to be multi-talented, fully aware of their field, and most importantly, fully aware of the interpersonal skills that yield successful organizational change. According to Heterick (1996), past president of EDUCOM, the information age demands, and will enforce, a transition to empowered employees throughout the organization. The organization will be successful to the extent that those employees are informed, free to exercise leadership, and capable of doing so.

Information resources and services leaders often find that the very characteristics that make their organizations capable of meeting the technological requirements of the campus serve to place them at odds with the institution's culture and values. For the information resources and services organization to assume a position of leadership on campus, the community must have trust in that group. Barone (1996), a technologist at the University of California, explains that trust is seeded through specific changes in traditional IT characteristics. Previous customer attitudes and perceptions of IT as being "controlling" are changed to "enabling", and the image of "provider" is shifted to "partner", while IT goals of "internal efficiency" evolve to "customer satisfaction".

The key to effective leadership, writes Battin (1996), who provided organizational leadership at both Columbia University and Emory University, is balancing the creative tension between the simultaneous capacity for decentralization and the requirement for central coordination to ensure broad and unencumbered access. McClintock (1997) contends that this balance is only achieved when higher education decision-makers recognize the implications of the information era, when they understand the importance of technology within the new educational paradigm, and when they know how to utilize information resources to their institutions' benefit.

Based upon his professional experience, Hirshon (1997) has determined certain institutional and organizational conditions that seem to predispose the leader toward success. He identifies one key condition as being strong support from the president of the institution.

DeMauro (1997), a technologist at Clarion University of Pennsylvania, suggests that strong support comes from simple communication. In the shadow of decreasing budgets and increasing demands, she attributes successful organizational change to open communication that breeds awareness of mission-critical processes and operations. Through this awareness, organizational change is more readily embraced.

Another factor in the leadership of successful organizational change seems to be the commitment to personnel development. Members of the professional staff need to feel that they are trained participants, and, thus, partners in the change process. LeDuc (1996b) states:

> Organizations need to ensure that there are backups, successors, and cross-training in order to guarantee organizational continuity. There is nothing so damaging to organizational health and the change process than to have someone leave who has become "irreplaceable". Enlightened leaders recognize the need to be protected from "too-critical" personnel by allowing enough time and attention to be paid to the issue of secondary support (p. 6).

As more colleges and universities opt for a closer relationship between computing and library organizations, much attention has been paid to the issues of organizational structure and control, funding sources and implications, and organizational benefits that can be derived from such a merger. Bortz (1993), a senior information administrator at George Washington University, maintains that leaders of organizational change in information resources and services must seek synergy of technology and organizational process. Merely pursuing technological change is not enough; leaders must implement a culture of change if they are to be successful in transforming the organization.

Likewise, Smith and West (1995) caution that very little mention is made of the cultural values that each group brings to the partnership and how those cultural values can either derail the merger or enhance it. Successful organizational change must promote a recognition of the positive aspects of both cultures. To exemplify these different values, Smith and West write:

> The language that the two groups use to essentially describe the same event is reflective of their two cultures. Computing language tends to be aggressive and war-like. Librarians tend to use words which emphasize harmony and group processes. Examples include: computing uses "end-user"; librarians use "patron". Computing says "execute routine and terminate processes"; libraries "lend and borrow". Computing "hacks a program"; librarians "search for information" (p. 6).

Bleed and McClure (1995) agree, conceding that information resources and services leaders need to be as familiar with the overall institutional culture as they are with individual departmental or professional cultures. Leaders must appreciate the importance of all existing cultures. By doing so, they will be determining the success or failure of organizational change initiatives that they must then facilitate and support.

Successful organizational change is also a function of being cognizant of the degree of customer understanding regarding those changes. Writing about organizational reengineering at Bryant College in Smithfield, Rhode Island, DeNoia (1993) asserts that the key to success is to reduce associated personnel, process, and organizational uncertainties. By emphasizing the enhancement of customer services – not the information and the technology itself – leaders of information resources and services integration projects will build credibility and forge lasting alliances among themselves and with the campus community. Pecka (1993), an administrator at Whitworth College in Spokane, Washington, agrees with increasing customer understanding and contends that users should be asked to provide input and suggestions that can be used to enhance their understanding of and ability to utilize information resources and services.

Having long been a field that tries to cope with rampant change, information technology and information service organizations acknowledge that rising customer expectations are the daily norm. Massey and Stedman (1995) warn that successful organizational change is subject to derailment if leaders neglect the effects of stress and the emotional health of those staff supporting the information needs of their customers. Increasing user demands, coupled with the loss of operating revenue and personnel resources, create a prolific environment for job-related stress. MacKnight (1995), suggests that when human resources are scarce, institutions should limit the scope of information services. To protect the emotional health of its staff, the organization must know and accept its own limitations. Hofstetter and Mullin (1993) contend that this understanding of organizational limitations must be realized before beginning any process aimed at increasing the effectiveness of customer service.

Integrating information technologies on campus raises complex issues and challenges. Gregorian, Hawkins and Taylor (1993), senior administrators at Brown University, caution that successful organizational change is predicated, not only upon the leaders' responsibility to offer access to enormous amounts of information, but also in their ability to assist in its transformation into knowledge. They write:

> Universities, colleges, libraries, and learned societies more than ever before have a fundamental historical and social task of giving users not merely training, but education; not only

education, but culture as well; and not just information, but its distillation – knowledge – to protect our society against counterfeit information disguising itself as knowledge (p. 2).

Further readings note that strategic planning by leaders is an impetus for successful organizational change. Rocheleau (1996), a member of the Northern Illinois University faculty, believes that, as expectations concerning information technology continue to soar, colleges and universities need to put more effort into linking their limited resources to bottom-line results through concerted strategic planning practices. Similarly, Ringle and Updegrove (1997), technologists at Reed College and Yale University respectively, write that many information technology leaders mistakenly view "technology planning" as an oxymoron. They caution that failing to recognize the need for technology planning can result in a great deal of wasted time, energy, and resources.

Many experts feel successful organizational change that promulgates the integration of computing and library services is predicated upon a strategy of customer-centered collaboration embraced by the information technology staff members and the library staff members. Katz and West (1993) conclude that interdependencies among administrative units should be identified and that the linkages between them should be strengthened. Fleit (1994), as president of EDUTECH International, notes that the institution should, in its planning, recognize the need for management and linkages between libraries, academic computing resources, administrative computing resources, telecommunications networking, and other learning centers. The chief information officer, writes Dolence, Douglas and Penrod (1990), is the senior-level administrator who should provide the leadership to all of these campus information resource and technology service and support areas.

McClure, Sitko and Smith (1997) also espouse collaborative internal relationships as critical components of successful organizational change. They conclude that in order to facilitate the transformations occurring in information resources and services, service providers must partner with faculty and administrators who possess the content and functional expertise. McClure, Smith, Stager and Williams (1993) further stipulate that the success of these collaborative efforts should be evaluated in terms of their effects on the academic process, not in terms of the state of technology. Bunker and Horgan (1994), senior administrators at the University of Washington and Seattle University, respectively, assert that chances for successful collaborative projects between computing and library services are increased if certain strategies are identified and employed from the outset; strategies like:

Focusing on customer needs first; getting top management support; developing and articulating joint mission statements and service agreements; planning for change and continuous process improvement; building flexibility into organizational structures; benchmarking services and processes with peers; outsourcing where needed to expand resources; negotiating ways to achieve synergistic effects; starting with small projects before moving to more ambitious collaborative efforts (p. 10).

Hardesty (1998) contends that the key to the leadership of effective organizational change is not found in the structure, but rather in the people involved – especially at small institutions with fewer personnel involved in the integration of resources and services. Battin and Hawkins (1997) concede that the present shape of college organizational structures and their somewhat myopic views are doomed. Battin and Hawkins believe that collaborative roles are evolving, and that these roles must be understood and accepted. In describing the successful leadership of this different kind of structure and approach, they write:

> Rather than advocating a specific organizational direction, it is more important to emphasize that there are different roles emerging for the professionals who try to integrate technology services that support faculty and student scholars on our campuses and provide the information they need (p. 1).

Perhaps one of the most important measures of the successful leadership of organizational change is the translation of information literacy to successful student outcomes. Academia has long required students to possess a prescribed level of basic reading and writing skills. Often a part of general education curricula, these skills ensure student success within and beyond the classroom. In today's learning environment, information literacy and information competency skills are becoming increasingly important.

Librarians traditionally have been responsible for teaching library research skills through a variety of techniques. However, these traditional approaches need to be rethought as educators face the task of teaching students to use and evaluate information in a burgeoning electronic environment. Goetsch and Kaufman (1997), administrators at the University of Maryland and the University of Tennessee respectively, identify a progression:

> Library instruction to bibliographic instruction to information literacy to information competency; as information resources and the tools to find them grow and increase in complexity, so, too, do the means by which we help our students understand how to find and use them effectively. What was once the province of librarians – teaching students how to use the library – now falls squarely on the shoulders of all of us: librarians, information technologists, instruction specialists, and faculty. Only by working collaboratively can we achieve our goals of educating students for the information age of the twenty-first century (p. 6).

Finally, a certain factor in the successful leadership of any organizational change is an understanding of the components of a human resources development system necessary to serve as the vehicle for change. Groff (1994, p. 31) has compiled the components of a technology-centered human resources development system into a definitive representation (see Figure 6).

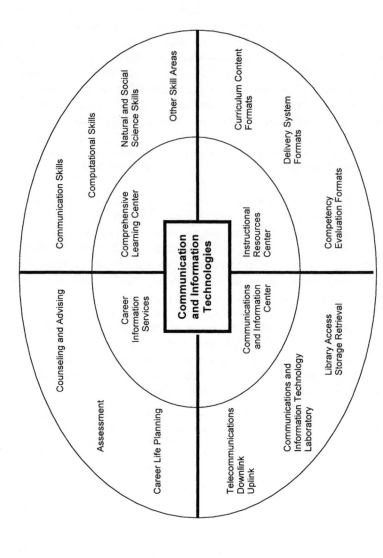

Fig. 6. Adapted from Groff, W. H. (1994). *Toward the 21st Century: Preparing proactive visionary transformational leaders for building learning communities.* Fort Lauderdale, FL: Nova Southeastern University. (ERIC Document Reproduction Service No. 372 239), p. 31.

Groff meticulously reveals how the skill sets and competencies that define desired outcomes are the products of a collaborative union between educational instruction and technological delivery. Components close to the core provide direction and focus to those necessary components lying further away from the center; yet, collectively they form a circle, or wheel suggesting forward progress. Kim (1998), in his own interpretation adopted from Overlock (1995), Branson (1990), and Anglin (1996), describes a similar technology-based paradigm (see Figure 7).

Kim clearly shows that achieving desired student outcomes now means making the transition from a teacher-centered, theory-based system to one that is student-centered, technology-based.

The significance of the problem at Nichols College was both considerable and disconcerting. Research indicated that this problem was not unique to Nichols College, but rather, was being addressed at similar institutions nation-wide. Clearly, many experts in the field of information resources and services were in agreement that the traditional organizational structures that define computing services and library services are changing rapidly because of the growing need and complexity of the technology that delivers users the information they require. Also fueling this transformation are the changing service and support demands and expectations of the campus community for whom the information resources and services are provided.

The analysis of the expert literature on the subject of organizational integration accomplished at similar institutions revealed that during the last five years, many academic institutions, small and large, have accepted the integration of services as a strategy for success. However, an understanding of cultural differences between computer services and library services was a key element for that success. Furthermore, the expert literature demonstrated that the demand for technology in academic libraries appears to be both a catalyst and an impetus for organizational change. This technology must work in concert with the academic curriculum to develop and enhance student information literacy skills. Finally, the expert literature reflected many sound characteristics that are viewed as pillars of leadership supporting successful organizational change. Also discovered was the significance of a human resources development system in a technology-based educational delivery system.

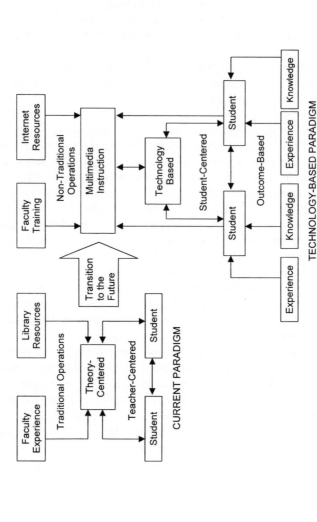

Fig. 7. Adapted from Kim, Y. G. (1998). A strategic plan for the integration of technology into the elementary teacher education program at Inchon University of Education. Unpublished doctoral dissertation, Nova Southeastern University, Fort Lauderdale, FL.

METHODOLOGY AND PROCEDURES

Assumptions

First, it was assumed that the search of expert literature was both complete and accurate. Second, it was assumed that the formative, critique, and summative committee members possessed the professional and educational knowledge and experience necessary to help guide the development of a strategic plan for the integration of computer services and library services at Nichols College. Third, it was assumed that the criteria developed by the formative committee and validated by the summative committee were an accurate reflection of the issues, constraints, considerations, competencies, skills, and cultural understandings needed for inclusion in such a strategic plan for organizational integration of information resources and services.

The development problem-solving methodology was utilized for this project. A strategic plan for the integration of computer services and library services to form a fully-integrated information resources and services group at Nichols College was the product developed from this research project.

Review of Expert Literature

Six procedures were used to complete this project and to answer the four research questions. First, a literature review was conducted to see: (a) how organizational integration accomplished at similar institutions, (b) how the demand for technology has grown in academic libraries, and (c) what characterized the leadership of successful organizational change. Furthermore, topics specific to information literacy, information competency, and the cultural difference between computer services staff and library services staff were also reviewed. The literature review also included an information search for models of organizational integration implemented at other colleges and universities and identified the components of a human resources development system. This review of expert literature assisted in answering the first two research questions: (a) "What are the practical benefits of organizational integration?" and (b) "What cultural concerns and differences affect the organizational integration?"

Analysis of Information

The second procedure for this project was to analyze information in order to understand and define how the effective delivery of integrated information resources and services could enhance the competencies and skills of the Nichols College student. These skills, known collectively as information literacy, were defined as the successful outcome of achieving and utilizing the cognitive competencies and technical skills necessary to access and analyze data from information resources with the intent of discovering knowledge that can then be used for planning, managing, and evaluation. Furthermore, these competencies, known collectively as information competency, were defined to represent the successful outcome of achieving and utilizing the cognitive competencies and technical skills necessary to interpret and extrapolate data from information resources with the intent of discovering knowledge that can then be used for planning, managing, and evaluation.

To assist in these and other analyses, a formative committee with seven members was established. These individuals represented faculty, computer services, and library services from both internal and external academic environments. The Management Information Systems (MIS) curriculum was utilized for the determination of desired student outcomes. It was believed to be imperative that such an understanding be built into the conceptual framework of the strategic plan in order for other institutional curriculums at Nichols College to also benefit from the integration of services. To accomplish this understanding, the current mission, goals, and objectives of the MIS department were analyzed to serve as a benchmark for the strategic plan. The skills associated with currently desired student outcomes of the MIS curriculum would serve as a map for success. The skills that students must possess included functional computer skills (including the PC-based operations of word processing, spreadsheet manipulation, and database structures) and networking skills (including Internet and World Wide Web access and search engine utilization). The competencies that students must possess include the ability to know which information databases will yield positive results, the ability to realize which data are relevant and which are not, and the ability to effectively search these databases through a concise understanding of the use of search operands and other data mining techniques.

The roles and functions of support staff necessary to achieve these desired student outcomes were also analyzed. In particular, the role of the electronic resource librarian (ERL) was analyzed in great depth. With the ERL possessing high information literacy skills and information competencies mandated by the electronic resources profession, recommendations from the ERL help adjust student competencies and skills in an effort to keep desired student learning outcomes in alignment with the current state of available information resources as well as the current level of user expectations and demands.

Models of organizational integration from other colleges and universities were analyzed. In order to better determine an appropriate model for reorganization and integration at Nichols College, overviews of expert literature gained from the responses of 97 institutions identified as having integrated computing and library operations were reviewed, and a comparative analysis of reporting models from the various institutions was made. This in-depth critique of the integration of information resources and services accomplished elsewhere helped to identify the practical, or functional benefits that could be anticipated at Nichols College. Furthermore, these organizational analyses served to capture the cultural concerns and differences found between traditional technologists and librarians that can impact the success of such an organizational merger. The goal of this critical analysis of external organizational structures was that a single model would emerge as a viable structure for Nichols.

The formative committee spent substantial time performing an analysis of this pertinent information. This second procedure assisted in answering the first two research questions: (a) "What are the practical benefits of organizational integration?" and (b) "What cultural concerns and differences affect the organizational integration?"

Refinement of an Organizational Model

The third procedure was to further refine the organizational model that emerged as a result of the second procedure. To accomplish this, the formative committee met on a weekly basis. The formative committee analyzed the expert literature presented for their consideration and documented the practical benefits of organizational restructuring at Nichols College. As the model for proposed organizational change at Nichols College took shape, the formative committee identified numerous internal processes requiring reengineering. These and other items of significance were documented in recorded minutes of the formative committee's weekly meetings.

Hoping that the mission, goals, and objectives of the MIS department would provide the framework for a successful student outcome, the formative committee defined criteria for a strategic plan for the integration of computer services and library services at Nichols College. The criteria satisfied all issues and concerns identified regarding practical requirements, cultural diversities, and process reengineering. This third procedure assisted in answering the third research question:

"What internal processes require reengineering to promote successful restructuring?"

Creation and Review of an Initial Draft

The fourth procedure was the creation of an initial draft of a strategic plan for the integration of computer services and library services at Nichols College. The initial draft of the strategic plan included, but was not limited to, an introduction, an executive summary of similar organizational structures, a detailed description of the organizational structure being proposed, the practical or functional benefits anticipated at Nichols College, the cultural concerns impacting a successful merger, the processes requiring reengineering, and an explanation of how knowledge management brought about by the effective delivery of information resources and services would support and enhance information literacy and information competency as student outcomes of the Nichols College curricula. The initial draft of the strategic plan provided recommendations for a timeline for implementation of the strategic plan.

Furthermore, as a part of the discussion of processes requiring reengineering, the initial draft of a strategic plan for the integration of computer services and library services described changes to the physical aspects of the building deemed appropriate to support the integration of services. The building that housed the former computer services, library services, and telecommunications groups promoted a physical layout that supported the distinct separation of these services departments. Physical changes to the building were required in order to accommodate the reengineering of information processes. The formative committee was cognizant of the Americans with Disabilities Act (ADA) regulations prior to these discussions, so that ADA considerations were incorporated into these physical changes to the building.

This draft was then distributed to a critique committee comprised of faculty and staff outside the formative committee, but internal to Nichols College. The critique committee reviewed the initial draft and provided critique and comments. The formative committee assessed comments made by the critique committee and incorporated those changes deemed appropriate into the document. After the draft passed between the formative committee and the critique committee, the issues and concerns voiced by the critique committee were summarized and documented. This fourth procedure assisted in answering the project's fourth research question: "What are the specific steps for a successful organizational integration?"

Validation of the Strategic Plan

The fifth procedure required that the formative committee present a draft of a strategic plan to a summative committee for expert validation and approval.

This summative committee of seven experts was comprised of individuals in key leadership positions both on and off the Nichols College campus.

The formative committee then asked that the Nichols College chief information officer (CIO) present the strategic plan to the summative committee for their expert review. This fifth procedure assisted further in answering the project's fourth research question: "What are the specific steps for a successful organizational integration?"

Revision and Submission of the Strategic Plan

The sixth procedure required that any revisions recommended by the summative committee be incorporated into a final draft of a strategic plan. All research questions were answered. The final strategic plan for the integration of computer services and library services to form a fully-integrated information resources and services group at Nichols College was submitted to the office of the president of the college. The formative committee remained available for any further discussion or clarification that the president might deem appropriate. It was recommended to the president that the strategic plan be presented to the Nichols College Board of Trustees for their information.

Limitations

This research project was limited in that it developed a strategic plan for the integration of computing services and library services at Nichols College based upon current information regarding models of organization. This limits the project in that, over time, Nichols College will need to again reorganize based upon changes in customer expectations and demands. In addition, due to the limited number of members available to participate on the formative and summative committees, the strategic plan reflects those committee member's perspectives on the integration of computer services and library services at Nichols College. The product developed as a result of this endeavor was limited in that it is intentionally specific to the concerns and considerations inherent to Nichols College. This product is not necessarily applicable in other settings.

RESULTS

The result of this research project was that a strategic plan for the integration of computer and library services at Nichols College was developed. The strategic plan (see Appendix D) addressed both short and long-range issues associated with the integration of information resources and services within the Nichols College learning community. The specific results of each of the six procedures used in the development of the strategic plan were reported.

Organizational Integration Models

The review of the literature provided critical insight into the project by assisting the formative committee in identifying by name other academic institutions that had gone through an integration of information services. These institutions were contacted for their expert opinions. Utilizing a list compiled by Hirshon (1998, pp. 35–37) of institutions that had integrated computing and library operations, a personal letter was sent via e-mail the second week of January 1999, to 97 institutions. During the course of the four weeks that followed this mailing, 51 or 52.6% responded. During the second week of March 1999, the same letters requesting library and technology organizational structure information were e-mailed to each of the 46 academic institutions who at that point had failed to respond to the first mailing. During the course of the next four weeks an additional 14 schools responded, bringing the total number of responses to 65, for a 67% rate of response.

Of the 65 colleges and universities that did respond, 17 of these institutions or 17.5% of the total number of schools listed are no longer operating under an integrated organizational structure. Of those 17 respondents, seven wrote that integrated services were tried for a time, and then for one reason or another those services were disintegrated. Four of those 17 respondents wrote that, while they had considered integration over the years, they had never actually implemented any reorganization that resulted in integrated computer and library services. Four of those 17 respondents wrote that while both groups reported to a single leader, they had in fact not integrated any of the collective roles, functions, or services of the two groups. And finally, two of those 17 respondents wrote that no such integration of computer services and library services existed at their institutions.

Of the 48 academic institutions remaining, and whose response indicated the existence of integrated computer and library services, 38 provided their organizational structures in response to the request made in the written mailings. Those organizational structures were made available to each member of the formative committee for comparative analysis and review in order to better determine the appropriate model for reorganization and integration at Nichols College (See Appendix A).

Technology Demand in Libraries

Literature from experts, both internal and external to the academic learning environment, was reviewed. This review of the literature provided critical insight to the research project by assisting in identifying the demand for information technology in academic libraries. It was revealed that the services offered by academic libraries have now reached outside the traditional walls of the library building and have landed on the desktop computer of the campus learning community. Because of both the capabilities and pervasiveness of technology, the constraints of time and place are fast being eliminated in the educational process. And this demand for technology in libraries has become a catalyst for change. From the creation of electronic reserves to web-based interlibrary loan (ILL) requests, patrons are coming to libraries with the expectation that technology is available that will accommodate their information needs.

Furthermore, the review of the literature revealed that this demand for technology in libraries is perhaps the greatest factor responsible for the synergy currently seen between computing services and library services. With the ever-growing electronic availability of information on both internal and external networks, many libraries have turned their focus to providing access upon demand rather than to building collections. In seeking to enable customers to locate information, librarians are stressing the need to provide resources and services to campus constituents upon demand using all forms of technology.

According to experts, this demand for technology in libraries is forcing a collaboration between computer services and library services never before seen. Many institutions, in a strategic move to solidify this collaboration, are now integrating these services through an organizational restructuring. By bringing about a marriage between the two functions, it is believed that a single voice addressing information and technology issues carries a greater value to the learning community than do two separate voices that often promote separate agendas.

Leadership of Organizational Change

This review of the literature provided critical insight to the research project by assisting in identifying concepts and strategies associated with the leadership of successful organizational change in the post-secondary educational environment. It was revealed that the most effective organizational changes occur when the leadership emphasizes the enhancement of customer services, as opposed to placing emphasis on the information or the technology. Time and again, the literature supported the belief that leaders of organizational change build credibility and forge lasting success when those who are affected by organizational change are dealt with in a supportive and collaborative manner.

Furthermore, this review of the literature revealed that the understanding of the components of a human resources development system becomes a certain factor in the successful leadership of organizational change. These components serve as the vehicle for change, and reveal how the skill sets and competencies that define desired student outcomes are the products of a collaborative union between the elements of educational instruction and technological delivery.

Analysis of Information

A formative committee was established to analyze the numerous dimensions associated with the project. These individuals were selected because of their participation in the educational process, as well as their knowledge of both the computing service and library service environment at Nichols College. The formative committee met on a scheduled basis from March 11, 1999, through June 17, 1999. The minutes of those meetings were recorded. The formative committee, in accepting its responsibilities, analyzed expert literature, performed a comparative analysis and review of external organizational structures, discussed practical benefits of organizational restructuring, identified internal processes that might require reengineering, and defined criteria for the strategic plan for the integration of computer services and library services at Nichols College.

Analysis of the Management Information Systems Curriculum

The second procedure for this project was to understand and define how the effective delivery of integrated information resources and services can enhance the skills and competencies of the Nichols College student. The Management Information Systems (MIS) curriculum was presented for the determination of these student outcomes. It was believed that such an understanding should be built into the conceptual framework of the strategic plan in order for other institutional curriculums at Nichols College to also benefit from the integration of services.

To accomplish this understanding, it was believed that the current mission, goals, and objectives of the MIS department should be analyzed in order to serve as benchmarks for the strategic plan. The formative committee learned that there was not a written mission statement for the MIS department at Nichols College. It was learned, however, that the goal of the MIS curriculum at Nichols College was to provide an extensive examination of the process by which pertinent and accurate information can be generated for use by the contemporary manager.

The formative committee learned that the objective of the MIS curriculum was to provide an understanding of the successful union between information systems and the decision-making process. To accomplish this, the MIS curriculum is comprised of two distinct curriculum components. The first is a systems analysis component (including programming) in which emphasis is on the design, analysis, and improvement of computer information systems. The second is a decision support component in which emphasis is on the analytical tools and techniques that underlie the decision-making process. Students emerging from this program of study are expected to be equipped to enter systems-related positions such as programmer/analyst, as well as other managerial positions that include objective decision-making as a key ingredient. The required courses for students enrolled in the MIS program of study were also identified.

The formative committee agreed that the skills associated with the desired student outcomes of the MIS curriculum should include functional computer skills (including the PC-based operations of word processing, spreadsheet manipulation, and database structures) and networking skills (including Internet and World Wide Web access and search engine utilization). The formative committee also agreed that the competencies that students must possess include the ability to know which information databases will yield the most positive results, the ability to realize which data are relevant and which are not, and the ability to efficiently and effectively search these databases through a concise understanding of the use of search engines, advanced search operands, and other data mining techniques.

However, it was the consensus of the formative committee that the reorganization and integration of computer services and library services could not in and of itself guarantee the enhancement of the skills and competencies of the Nichols College MIS student. After much formative committee discussion, it was decided that faculty must first create the demand for information resources and services, rather than creating an expectation that the supply of information resources and services would ensure successful student outcomes. There was committee agreement that inadequate and inefficient information resources and services can impede successful student outcomes, but that even the best information resources and services have little if any effect on the skills and competencies of student if faculty are unwilling to require students to utilize the resources.

The formative committee did agree that a newly-created Information Resources and Services group should seek to achieve specific departmental goals that would create an environment that is more efficient and supportive of the faculty efforts to integrate technology into curriculum and cultivate successful student outcomes. The formative committee believed that these departmental goals should become the essence of the strategic plan for the integration of computer and library services.

Analysis of Roles and Functions of Support Staff

The role and functions of support staff necessary to support desired student outcomes were also analyzed. In particular, the formative committee sought to review in detail the role of the electronic resource librarian (ERL). The formative committee believed that with the ERL possessing high information literacy skills and information competencies mandated by the electronic resources profession, recommendations from the ERL would help adjust student skills and competencies in an effort to keep desired student outcomes in alignment with the current state of available information resources.

After much discussion, it became the consensus of the formative committee that the roles of those individuals who work in Information Resources and Services (IRS) remain in fact unchanged, whether or not computer services and library services were merged. However, the "official" job description for each IRS member was out of date, and it was agreed that new job descriptions should be written that would capture the roles as perceived in today's information world. Each member of the IRS group agreed to assist in the rewriting of existing job descriptions that were years old and to forward these revised job descriptions to the chief information officer (CIO) at Nichols College for review and approval.

Within two weeks following this formative committee request, it was verified by the formative committee that the job descriptions had been rewritten by the Information Resources and Services group members, had been reviewed and approved by the CIO, and had been submitted to the Human Resources Department. With the approval of the president of the college, an organizational structure describing the proposed Information Resources and Services group was also provided to the Human Resources Department at that time. The director of the Human Resources Department agreed to provide feedback to the formative committee if the need arose.

The formative committee did agree that the electronic resource librarian (ERL) would assume a key role in the collaborative union between the faculty, the students, and the available information resources and services. While the faculty would be expected to create the demand for information resources and services, and while the IRS group would be expected to supply the necessary information resources and services, it would be the ERL who would be expected to bring the two together in a cogent and meaningful way that would facilitate individual learning.

Analysis of Models of Organizational Integration

The formative committee made a comparative analysis of reporting models from 38 institutions in order to better determine an appropriate model for reorganization and integration at Nichols College. This in-depth critique of integrations of information resources and services accomplished elsewhere helped to identify the practical, or functional benefits that could be anticipated at Nichols College. Benefits identified through this formative committee analysis included centralized acquisitions (purchasing) for computer services and library services, centralized audio/visual services, centralized help desk operations, a combined campus technology newsletter representing both computer services and library services, centralized administrative functions, and a centralized computing facility for the access of information on demand.

Furthermore, these organizational analyses by the formative committee served to capture the cultural concerns and differences found between traditional technologists and librarians that can impact the success of such an organizational merger. Figures 1–3 present the underlying perceptions that contribute to the cultural differences found between many computer services and library services staff members.

Through this critical analysis of external organizational structures by the formative committee, a single model emerged as the most viable structure for Nichols College to adopt. The formative committee agreed that the new organizational model of an integrated structure seemed best suited for the Nichols College learning community (see Appendix B). It was also agreed that the reorganized structure, which reflected the merging of the Information Technology department, the Telecommunications department, and the Library department, would retain the name Information Resources and Services. The formative committee supported the premise that all three departments would report to the chief information officer of the college.

Refinement of an Organizational Model

The formative committee continued to meet on a scheduled basis in an effort to further define the organizational model that emerged as a result of the second procedure. This refinement of the organizational model assisted in answering the third research question. From the formative committee's analysis of the expert literature presented for their consideration, the practical benefits of organizational restructuring at Nichols College were clearly identified. These benefits included centralized acquisitions (purchasing) for computer services and library services, centralized audio/visual services, centralized help desk operations, a combined campus technology newsletter representing both computer services and library services, centralized administrative functions, and a centralized computing facility for the access of information on demand.

Internal Processes Requiring Reengineering

As the model for proposed organizational change at Nichols College took shape, the formative committee identified numerous internal processes requiring reengineering. These and other items of significance were documented in recorded minutes of the weekly meetings.

Of all internal processes identified by the formative committee as possible candidates for reengineering, changes to the physical facilities became the initial focus of the committee. The current physical description of the library building was documented, and the committee spent time discussing changes that would result from the integration of computer services and library services. It was agreed that the existing physical environment promoted the distinct separation of services, and consequently, the committee focused on how the separate groups could be repositioned in a manner that would promote integration and collaboration.

The formative committee agreed that the office of the chief information officer should be relocated from the first floor of the library building to the third, or "main" floor of the building (since the building is located on the side of a hill, the main public access on the front of the building enters at the third floor level of the building). It was also agreed that the office of the director of academic computing should be relocated from the first floor of the building to the third, or "main" floor of the library. Also remaining on the third floor of the building would be the offices of the electronic resource librarian, the patron services coordinator, the serials and technical services coordinator, the acquisitions clerk, and the evening library supervisor. It was agreed by the committee that the office of the director of library should be relocated from the third floor to the fourth floor of the building.

Remaining on the first floor of the library building within the Information Technology department would be the offices of the director of administrative computing, the network administrator, the system analyst, and the computing and network technician. Also to remain located on the first floor of the library building within the Telecommunications department would be the office of the telecommunications manager.

Next, the formative committee identified and discussed numerous functional issues and considerations that led to internal processes requiring reengineering. Among these, the committee identified the closing of the computer lab on the first floor of the library building and the placement of those computers and printers throughout the remaining floors of the building, the location of laptop computer access jacks throughout the building, the logistics of a centralized technology help desk located at the circulation desk of the library, the required training associated with all reengineered processes, and the need to migrate from printed reserves to electronically-served reserves at Nichols College.

Definition of Criteria for the Strategic Plan

The formative committee spent considerable time together both in physical meetings and in virtual electronic discussions reviewing the expert literature relating to organizational integration models, technology demand in libraries, and the leadership of organizational change. In addition, the formative committee had numerous discussions involving the analysis of the MIS curriculum, as it related to the information resources and services available for the enhancement of the information literacy and information competency skills of the Nichols College learning community. The formative committee reviewed the roles and functions of support staff within the information resources and services environment and performed a comparative analysis of other organizational structures. The formative committee identified cultural differences between the two currently distinct service providers. Furthermore, the formative committee identified both physical and functional concerns that would require reengineering. With this research serving as the guide for the task of developing a strategic plan for the integration of computer services and library services at Nichols College, the formative committee next began the task of defining the criteria for the strategic plan.

The formative committee agreed that to be successful, the criteria must satisfy issues and concerns identified regarding practical requirements, cultural diversities, and process reengineering, as well as address the mission and goals of Information Resources and Services. In an effort to better understand the structure and criteria associated with a strategic plan related to computing and library services, the formative committee reviewed numerous strategic plans from other academic institutions. With assistance provided by the chief information officer, the formative committee was directed to the EDUCAUSE website. EDUCAUSE maintained at the time an information resources library with over 140 strategic plans from other academic institutions at http://www.educause.edu/ ir/library/ subjects/plans-strategic.html[.] In addition, the chief information officer provided copies of sample strategic plans to the formative committee for their consideration and review.

After several discussions, the formative committee established the criteria for a strategic plan for the integration of computer services and library services at Nichols College. This criteria (see Appendix C) addressed the issues and concerns previously identified by the formative committee. The following criteria were established by the formative committee:

(1) The plan would include a statement of introduction.
(2) The plan would include a mission statement that represented the Information Resources and Services group within the context of the Nichols College mission.
(3) The plan would include a statement of history of the past technological and informational environment.
(4) The plan would include a description of the current information resources and services environment in terms of the physical, organizational, functional, and cultural components.
(5) The plan would include recommendations for the future information resources and services environment in terms of the physical, organizational, functional, and cultural components.
(6) The plan would include short term (one-year) goals of the Information Resources and Services group.
(7) The plan would include long term (two to five-year) goals of the Information Resources and Services group.
(8) The plan would include a mechanism for the annual review and assessment of the strategic plan goals.
(9) The plan would include recommendations for implementation, including a timeline for implementation.
(10) The plan would be succinctly written in simple, non-technical English.
(11) The plan would be made available on the Nichols College website for public access.

The chief information officer recommended to the formative committee that an executive summary of similar organizational structures be incorporated into the strategic plan. The formative committee, however, decided that an executive summary within the strategic plan was not desirable. The formative committee did agree, however, that an executive summary should be available within the Information Resources and Services department. The chief information officer created the executive summary, distributed it to the formative committee for review, and placed it within the Information Resources and Services department for future access and review.

Creation and Review of an Initial Draft

During the fourth procedure of the project, the formative committee created an initial draft of the strategic plan document and worked closely with the critique committee to produce an initial draft ready for summative committee validation, all of which assisted in answering the fourth research question: "What are the specific steps for a successful organizational integration?" The initial draft of the strategic plan addressed all the criteria established by the formative committee.

Creation of the Strategic Plan Draft

The initial draft of the strategic plan included the following components: (a) a statement of introduction; (b) a mission statement that represented the Information Resources and Services group within the context of the Nichols College mission; (c) a statement of history of the past technological and infor-mational environment; (d) a description of the current information resources and services environment in terms of the physical, organizational, functional, and cultural components; (e) recommendations for the future information resources and services environment in terms of the physical, organizational, functional, and cultural components; (f) short term (one-year) goals of the Information Resources and Services group; (g) long term (two to five-year) goals of the Information Resources and Services group; (h) a mechanism for the annual review and assessment of the strategic plan goals; and (i) recom-mendations for implementation, including a timeline for implementation. Furthermore, the initial draft strategic plan was succinctly written in simple, non-technical English. Included in the fifth component was a statement of belief relative to how knowledge management brought about by the effective delivery of information resources and services perpetuates an environment that supports and enhances information literacy and information competency as a student outcome of all Nichols College curricula.

Furthermore, as a part of the fifth component, the initial draft of a strategic plan for the integration of computer services and library services described changes to the physical aspects of the building deemed appropriate to support the integration of services. The building, which housed computer services, library services, and telecommunications, promoted a physical layout that supported the distinct separation of these service departments. Certain physical changes to the building were required in order to accommodate the reengi-neering of information processes.

Review of Draft Document by Critique Committee

After the development of the initial draft of the strategic plan for the integra-tion of computer services and library services at Nichols College, the formative committee sought expert input from a critique committee. Six individuals repre-senting faculty and staff comprised the critique committee. These individuals, while outside the formative committee, were internal to Nichols College. The formative committee asked that the chief information officer present the strategic plan to the critique committee for review.

A list of critique committee concerns was developed as a component of the detailed documentation of this research project. The critique committee comments were brought to the formative committee for discussion. With consideration for the input received from critique committee, the formative committee finalized a draft of the strategic plan (see Appendix D), and agreed that the strategic plan should be forwarded to the summative committee for review.

Validation of the Strategic Plan

After the development of the draft of the strategic plan for the integration of computer services and library services at Nichols College, the formative committee sought expert input through a summative committee review. This committee of experts was comprised of individuals in key leadership positions both on and off campus. The formative committee asked that the Nichols College chief information officer present the strategic plan to the summative committee for this high-level expert review.

The responses from individual summative committee members were varied. Of the seven summative committee members, five indicated approval of the draft strategic plan as written with no comments noted. One summative committee member indicated approval of the draft strategic plan as written with one comment noted. One summative committee member indicated approval of the draft strategic plan as written with ten comments noted.

The end result of the summative committee review was that the draft strategic plan was approved as written. The eleven total comments were presented to the formative committee for discussion. Those summative committee comments were:

(1) It might be helpful to adopt the plan with specific definitions of who is primarily responsible and who is contributing to the success of each activity, along with specific dates for completion of the activities.

(2) It might be helpful to define a mission statement for each of the functional areas of Information Resources and Services to include the Information Technology department, the Library department, and the Telecommunications department.

(3) It might be helpful if the library defined its strategic objectives for collection development, given that information utilities are allowing academic libraries to alter the fundamental composition of their traditional collection of books and journals to an enlarged cyber environment of integrated digital and media resources.

(4) Electronic collections now accent the library as a service point and de-emphasize the library's role as a place for collections. This shift in focus is pushing the library to become more service focused. It might be helpful if the library developed a plan regarding bibliographic instruction and the introduction of web-based tutorials, developed in partnership with faculty to support information competencies stemming from bibliographic objectives established by the faculty for their students.

(5) The implementation of electronic reserves is an innovation that would greatly benefit the Nichols College academic community. It would likely require a lot of time and energy. It would also require the active participation of library staff to work in academic partnership with classroom faculty. The use of electronic resources will undoubtedly set records compared to the circulation of print resources. It might be helpful if counters could be put on all web pages to track this and to follow the use of the library's web pages. Furthermore, it was believed that when the electronic reserve system is implemented, students may embrace the new technology with enthusiasm, only to then find that faculty continue to assign old out-of-date materials. It might also be helpful if the library developed a plan for working in academic partnership to update the resources that the Nichols College faculty use and assign.

(6) A built-in effect of the library's integrated information system is that it allows the clientele to see if the library's database infrastructure is weak. It might be helpful if the library developed a written plan that would address what the library is currently doing or plans to do in the future in terms of improving user services through shelf reading, collection inventory (in whole or of high-use areas), or targeted collection development and weeding.

(7) The Nichols College library is open approximately 84 hours per week. The availability of web-based resources with remote access could have the effect of lowering the per-transaction cost for library resources. It might be helpful for the Information Technology department to develop a plan that addressed how the remote access environment would be upgraded over time to accommodate the effects of increased enrollment, increased utilization, and hardware and software changes in the infrastructure technology.

(8) It might be helpful if the library developed a plan that addressed the state of the library's book collection and the possible impact of electronic books. It was already known that netLibrary (Boulder, Colorado) had launched a comprehensive digital book collection. Electronic book collections, like full-text journal collections, will allow libraries to better serve patrons by reducing or eliminating the costs associated with storage, acquisition/cataloging, replacement, loss, and maintenance.

(9) The strategic plan mentions a newsletter. It might be of greater impact to the Nichols College learning community if the objective were larger in scope than just the Information Resources and Services group. That is, it might be of greater value to all if the objective were for Nichols College to initiate a cyber journal instead. The focus of this web-based cyber title could be to publish the best academic work of Nichols's students. In doing so, it would teach something about publishing and would create a lot of pride and recognition for Nichols College.

(10) Technology support requires a sufficient technical staff. Generally, there is not enough money to hire enough full-time people, so the next best scenario is to use students. It might be helpful if each of the three functional areas of Information Resources and Services developed a plan that addressed the involvement of students in the technology program. The specifics of the student involvement should include help desk coverage, installation of software, the troubleshooting of problems, helping students who have problems with technology in the library, and such. Since a college is about teaching and the students are there to learn, there is an inherent synergy when students are involved in technology service and support.

(11) The strategic plan did not address copyright issues, and how Nichols College plans to comply with The Digital Millennium Act. If this has not already been addressed outside of the strategic plan, it might work well for the chief information officer to serve as the College Copyright Officer.

Revision and Submission of the Strategic Plan

The formative committee discussed the comments presented by the summative committee. The formative committee agreed that all were valid, constructive comments. Although outside the realm of the strategic plan, four of the summative committee's comments (comments 1, 7, 10, and 11) dealt with issues that had already been addressed by individual functional areas within the Information Resources and Services group. The formative committee agreed that the remaining seven comments should be addressed as thoroughly as possible.

The agreed upon method of addressing these remaining comments was to incorporate them into the annual assessment and review process described in the strategic plan. As new goals are added to the strategic plan each year, the opportunity would then exist for the helpful comments supplied by summative committee members to be transformed into strategic goals. Furthermore, the formative committee agreed that both the critique committee members and the summative committee members should be solicited for comment each year during the process of setting new goals for the Information Resources and Services group at Nichols College.

Based upon the willingness and indicated approval of the formative, critique, and summative committees involved in this research project, the draft strategic plan was deemed final. The chief information officer for Nichols College presented the document, entitled *Nichols College Information Resources and Services Strategic Plan 1999–2004*, to the president of Nichols College for approval. The chief information officer recommended that the strategic plan be approved, adopted, and implemented as soon as possible. It was also recommended to the president that the strategic plan for the integration of computer and library services be presented to the Nichols College Board of Trustees for their information.

DISCUSSION, CONCLUSIONS, IMPLICATIONS, AND RECOMMENDATIONS

Discussion

The result of this research project was that a strategic plan for the integration of computer services and library services at Nichols College was developed. The strategic plan (see Appendix D) addressed both short and long-range issues associated with the integration of information resources and services within the Nichols College learning community. The specific results of each of the six procedures used in the development of the strategic plan were reported. Those six procedures were: (a) review of expert literature; (b) analysis of information by a formative committee; (c) refinement of an organizational model; (d) creation of an initial draft of a strategic plan by a formative committee, with a review of that draft by a critique committee; (e) validation of the strategic plan by a summative committee; and (f) revision and submission of the final strategic plan.

Throughout the project, subject literature from experts was reviewed. This review provided critical insight to the project by assisting the formative committee in identifying the names of other academic institutions that had gone through an integration of information services. The formative committee sought these names so that other institutions could be contacted for their expert opinions regarding the integration of services.

The formative committee learned that many colleges and universities had begun to address the problem of elevating present levels of effectiveness in providing information resources and services. Some institutions have adopted the strategy of integration of library services and computer services as a cause for greater collaboration between information resource and service providers within the academic learning environment.

The strategic plan developed as a result of this research project was fashioned in a way that the use of current information technologies were promoted on behalf of Nichols College's instructional programs and academic curricula.

Conclusions

Ten conclusions were reached. These conclusions were as follows:

(1) The demand for technology in libraries is forcing collaboration between computer services and libraries services, and in a strategic move to solidify this collaboration, Nichols College should integrate these services through an organizational restructuring defined in the strategic plan.

(2) The recommended organizational structure defined in the strategic plan for Nichols College is one that addresses the concerns, challenges, and aspirations of each specific constituency.

(3) The integration of computer services and library services as defined in the strategic plan should foster an environment that recognizes, embraces, cultivates, and utilizes the varied staff cultures of the Information Resources and Services group.

(4) The integration of computer services and library services as defined in the strategic plan should foster an environment that explores and challenges out-dated functional processes, and accepts the tenet that the reengineering of operational processes is a healthy and desirable part of the professional workplace at Nichols College.

(5) Effective organizational changes can occur at Nichols College when the leadership emphasizes the enhancement of customer services, as opposed to placing emphasis on the information or the technology.

(6) The reorganization and integration of computer services and library services does not, in and of itself, guarantee the enhancement of the skills and competencies demanded of the Nichols College student.

(7) Inadequate and inefficient information resources and services at Nichols College can impede successful student outcomes, however the best information resources and services have little if any effect on the skills and competencies of student if teaching faculty are unwilling to require Nichols College students to utilize the resources.

(8) The establishment of goals within the Information Resources and Services group and as defined within the strategic plan, is a valid mechanism that will help ensure the professional growth of the entire Nichols College learning community.

(9) While none of the organizational structures studied as a part of this research project should be considered perfect, and none are likely to become terminally dysfunctional, it is the institutional culture of the institution and the willingness of the individuals involved to adopt new models of thinking that will determine the ultimate success of Information Resources and Services at Nichols College.

(10) The integration of computer services and library services as Information Resources and Services at Nichols College is an appropriate and desired organizational change, and the strategic plan developed as a result of this research is a viable mechanism toward that end.

Implications

The development of a strategic plan for the integration of computing services and library services occurred within the intent and spirit at Nichols that all information resources on campus to be integrated and aligned with its mission. This organizational change should increase the college's effectiveness in providing information resources and services. Without this strategic plan, it was believed that, given budgetary resources available, the library might continue to serve students by doing little more than acquiring and archiving information. However, through the successful integration of information resources and services, the library could evolve from being a place where information is kept to become a portal through which learning students and professionals could access the vast resources of the world on demand.

In responding to the pressure for greater information resources access, the organization needed to adopt a more consumer or customer service orientation. This strategic plan for integration of computer and library services within the Nichols College learning community should help codify successful processes and practices intrinsic to the institution and blend them with the best of currently emergent customer-oriented organizational models. This convergence of information and the technology upon which it relies should help students master a body of knowledge and develop greater critical thinking skills. This plan should promote the information literacy of Nichols College students by assisting learners in locating, extracting, and processing information in order to generate needed knowledge. Colleges and universities have a responsibility to society to develop strategic plans which can establish a solid foundation for the implementation, utilization, and delivery of needed educational opportunities. The thought and effort that went into the development of this strategic plan can ultimately define the success of information resources and services at Nichols College during the next decade.

Recommendations

Five recommendations germane to the integration of computer services and library services at Nichols College were made as a result of this research project:

(1) The president of Nichols College should approve as soon as possible the strategic plan that was the resulting product developed from this research project.

(2) The Information Resources and Services group at Nichols College should, at the beginning of the 1999–2000 academic year, adopt and implement the strategic plan that was the resulting product developed from this research project.

(3) In order for the strategic plan to remain a viable and effective document during the course of its proposed span, Information Resources and Services should, in conjunction with the senior administration, perform an annual assessment and review of the strategic plan.

(4) Due to the changing nature of the technologies associated with information access and retrieval, as well as the changing types of media available for information storage, the subject of the integration of computer services and library services in the academic environment should be researched as often as the user expectations and demands for this information change.

(5) Other academic institutions that have gone through an integration of information services, or who are considering some sort of integration of information services should have the opportunity to review this research, with the most likely mechanism for vast dissemination being the World Wide Web.

APPENDIX A

Current Models of Organization Integration

Of the 48 academic institutions whose response indicated the existence of integrated computer and library services, 38 of those institutions provided their organizational structures in response to the request made in the written mailings. The following organizational structures were made available to each member of the formative committee for comparative analysis and review in order to better determine the appropriate model for reorganization and integration at the Nichols College campus.

Name of Institution

Aurora University (IL)
California Lutheran University (CA)
California State University-Bakersfield (CA)
California State University-Chico (CA)
California State University-San Bernardino (CA)
Carthage College (WI)
Case Western Reserve University (OH)
Concordia University (CA)
Connecticut College (CT)
Eckerd College (FL)
Indiana State University (IN)
Kalamazoo College (MI)
Kenyon College (OH)
Lehigh University (PA)
Macalester College (MN)
Mankato State University (MN)
Michigan State University (MI)
Mount Holyoke College (MA)
Oregon State University (OR)
Rensselaer Polytechnic Institute (NY)
Saint Mary's College (MD)
Saint Michael's College (VT)
Southern Illinois University-Edwardsville (IL)
State University of New York-Brockport (NY)
State University of New York-Cortland (NY)
State University of New York-Potsdam (NY)
University of Rhode Island (RI)
University of Richmond (VA)
University of Scranton (PA)
University of South Carolina (SC)

University of Southern California-Los Angeles (CA
University of Utah (UT)
University of Wisconsin-Oshkosh (WI)
University of Wisconsin-Parkside (WI)
University of Wisconsin-Whitewater (WI)
Virginia Commonwealth University (VA)
Virginia Polytechnic Institute and State Univ. (VA)
William Paterson University (NJ)

APPENDIX B

Organizational Structure for Nichols College

The organizational structure shown in Fig. 8 was endorsed by the formative
committee for the Information Resources and Services group at Nichols College.

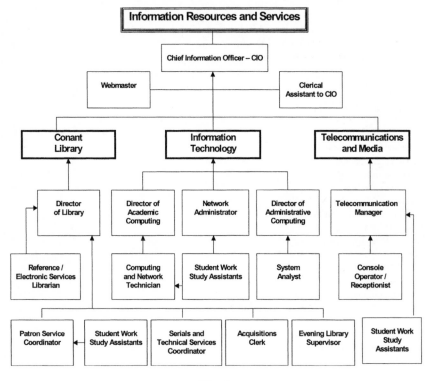

Fig. 8.

APPENDIX C

Criteria Developed for the Strategic Plan

As a result of the third procedure of this research project, the formative committee developed a list of criteria deemed both necessary and desirable. The formative committee utilized these criteria in the initial structuring of a strategic plan for the integration of computer services and library services at Nichols College. The following criteria were established by the formative committee:

(1) The plan would include a statement of introduction.
(2) The plan would include a mission statement that represented the Information Resources and Services group within the context of the Nichols College mission.
(3) The plan would include a statement of history of the past technological and informational environment.
(4) The plan would include a description of the current information resources and services environment in terms of the physical, organizational, functional, and cultural components.
(5) The plan would include recommendations for the future information resources and services environment in terms of the physical, organizational, functional, and cultural components.
(6) The plan would include short term (one-year) goals of the Information Resources and Services group.
(7) The plan would include long term (two to five-year) goals of the Information Resources and Services group.
(8) The plan would include a mechanism for the annual review and assessment of the strategic plan goals.
(9) The plan would include recommendations for implementation, including a timeline for implementation.
(10) The plan would be succinctly written in simple, non-technical English.
(11) The plan would be made available on the Nichols College website for public access.

APPENDIX D

Strategic Plan

The strategic plan developed by the formative committee included the following components: (a) a statement of introduction, (b) a mission statement that represents the Information Resources and Services group within the context of the Nichols College mission, (c) a statement of history of the past technological and informational environment, (d) a description of the current information resources and services environment in terms of the physical, organizational, functional, and cultural components, (e) recommendations for the future information resources and services environment in terms of the physical, organizational, functional, and cultural components, (f) short term (one-year) goals of the Information Resources and Services group, (g) long term (two to five-year) goals of the Information Resources and Services group, (h) a mechanism for the annual review and assessment of the strategic plan goals, and (i) recommendations for implementation, including a timeline for implementation of the plan.

Furthermore, the strategic plan was succinctly written in simple, non-technical English. The following is the final strategic plan, as presented by the chief information officer to the president, after having incorporated all concerns indicated by the formative committee, the critique committee, and the summative committee:

Nichols College
Information Resources and Services

STRATEGIC PLAN

1999–2004

Introduction

During the 1998–1999 academic year, the decision was made to integrate computer services, library services, and telecommunications into a more customer-oriented group named Information Resources and Services. The group, under the leadership of the chief information officer and with recommendations provided by a committee comprised of representatives from various campus constituencies, sought to define more effective utilization and growth of its information services to the Nichols College learning community. This strategic plan defines those directions such that the mission and goals of Information Resources and Services are identified.

Mission Statement

Information Resources and Services has adopted the following mission statement:

"The mission of Information Resources and Services is to support and enrich the Nichols College learning community. By providing instruction, information access, staff expertise, and technology, Information Resources and Services addresses the academic, informational, and instructional needs of Nichols College and the surrounding community".

The above mission of Information Resources and Services was written within the spirit and context of the Nichols College mission statement:

"Nichols College is committed to providing the best practically-oriented business education in New England. We transform students into competent, assertive business professionals, who respond to challenges, are eager for responsibility, and stand ready to rise in the world of commerce. The results-oriented education at Nichols is supported by comprehensive exposures to technology, the arts and sciences, a full range of athletic opportunities, and a broad spectrum of extracurricular activities. This mix produces graduates who are well-rounded, sophisticated, computer literate and prepared to take their places in the executive community. The business program at Nichols is paralleled by comprehensive programs in teacher education and the liberal arts. The Nichols experience happens in a friendly, nurturing small college environment where students are encouraged to invent themselves under the guidance of an experienced faculty and staff that stresses teaching excellence and concern for student welfare and success" (adopted April 17, 1999).

History

Information Resources and Services has evolved from a history of dedication to technological excellence at Nichols College. Nichols College, being one of the first colleges in America to require every student to purchase and bring a laptop computer into the classroom, has long been firm in its commitment to provide and maintain a positive information technology infrastructure to all who work and study there.

Nichols College completed the implementation of a high-speed fiber-optic voice and data infrastructure in 1997. This network was designed to support its newly acquired digital telecommunications system, voice mail, Internet access, electronic mail, and data environments well into the future. With this infrastructure available to all students, faculty, and staff, the ability for users to access information on demand was greatly enhanced. Also occurring in 1997, the Computer Services department at Nichols College was renamed Information Technology (IT). Rationale for this change followed the belief that technology has become so pervasive and so diverse, that the term "computer services" had become passé.

Over the years, the IT department became responsible for the complete workings and seamless integration of communications, computer systems, and the network infrastructure. IT was also responsible for the service and support required to back up each of those endeavors. Furthermore, IT served Nichols College by providing for the security, retention, backup, and recovery of electronic information and voice data. IT also served by maintaining automated information systems, and by providing training, technical consultation, and planning in the application of information technology in both the academic computing and administrative computing environments.

The academic library environment has been, and will continue to be, one of extraordinary change. The library has seen huge increases in the amount of scholarly publication, large increases in the cost of information resources – particularly in the area of print journals, and rapidly changing information technologies requiring frequent equipment upgrades and training. The library has seen shifts in traditional educational techniques and patterns, user expectations, and methods of information acquisition and distribution. Increasingly, the library is positioning to become a research partner rather than simply an information provider.

Consequently, the library at Nichols College has been a critical part of the history of the college. Since its beginning, the primary mission of the library has been to support and enrich the academic programs and teaching curriculums of Nichols College. During 1998, the library added a full-time electronic reference librarian to its staff, further confirming its commitment in the successful attainment of desired student outcomes. Today, the library offers a diverse array of materials for the intellectual, professional, and cultural development of the students, faculty, and staff. These materials, be they printed or electronic, are available to address the information needs of the entire Nichols College learning community.

Current Environment for Information Access and Support

Information Resources and Services represents the reorganization and merging of library services, computer services, and telecommunications services into a strategically aligned group sharing a unified vision and single mission. This new organization identifies information as the pillars of its operations. While building upon the strengths of all its members, this collaborative group views each individual as a strategic partner, and provides an environment that nurtures the opportunity for professional growth and the development of new skills.

Historically, the library services have focused upon the selection, acquisition, organization, description, automation, and preservation of information. Computing services have traditionally provided the hardware and software that emphasized the technology, with information seen as a secondary byproduct. Telecommunication services have always been concerned with the transfer of information from source to destination. Nichols College recognized that recent concentrations within the Information Technology department on reengineering various campus activities (with an emphasis on the retrieval of relevant data for various constituents) was congruent with the Library's client-centered and information-related mission. During that same time, the technology associated with the Telecommunications department had become compatible with the networked data infrastructure already in place. The demand for digital telephony including voice mail capability helped bring about a leading edge telecommunications environment for all students, faculty, and staff.

The Information Resources and Services group was created to capitalize upon this growing synergy, and encompasses the functional areas of computing services, library services, and telecommunications services. Physically located in the library building, the Information Resources and Services group has addressed both its new organizational structure and staff job descriptions with the Human Resources department at Nichols College. Immediately following the creation of this group, a new organizational chart was created, indicating that the three departments of Information Technology, Library, and Telecommunications now report to the chief information officer. Updated job descriptions were also created to better define the roles of the members of the Information Resources and Services group. The Information Resources and Services organizational structure is defined as follows (and as shown in Fig. 8).

The Information Resources and Services group functions as a collaborative team comprised of two distinct professional cultures. Those staff providing library services to Nichols College have similar professional backgrounds and library science education, while those staff providing technical services to Nichols College (Information Technology and Telecommunications) have dissimilar professional backgrounds and educational experiences. The success of the current group is achieved through the understanding of the similarity of roles in the college learning environment. Capitalizing upon these similar roles, the two dissimilar cultures work together to provide information access, integrated services, and information retrieval instruction.

Future Environment for Information Access and Support

Information Resources and Services is committed to its mission of support and enrichment to the Nichols College learning community. In order to achieve that mission, the technological and informational environment must be reviewed, planned, and changed at a pace that coincides with the need for both the personal and professional growth of its learning community.

This strategic plan identifies two major groups of goals that will support the mission of the Information Resources and Services group. The short term goals listed are recommended for completion during the 1999–2000 academic year. The long term goals are recommended for completion during the 2000–2001, 2001–2002, 2002–2003, and 2003–2004 academic years. Collectively, the goals established in this five-year strategic plan support the mission of Information Resources and Services, the mission of Nichols College, and the vision that the college president and the Nichols College board of directors have defined for the next millennium.

Goals for the 1999–2000 Academic Year

The short term goals identified for completion during the 1999–2000 academic year are defined as follows:

- The physical integration of library, information technology, and telecommunications staff should be completed. To date, the chief information officer has been relocated from the first floor to the third floor of the library building. The director of academic computing has been relocated from the first floor to the third floor of the library building. The director of the library has been relocated from the third floor to the fourth floor of the library building. The acquisitions clerk should be relocated from the third floor to the first floor of the library building. These physical relocations support the integration of informational processes, and encourage the integration of all cultures found within Information Resources and Services.
- The acquisitions clerk should become responsible for all aspects of purchasing, receiving, budget posting, invoice processing, and monthly budget reconciliation for the entire Information Resources and Services group. Currently, the chief information officer has assumed this responsibility on behalf of the Information Technology department and the Telecommunications department, while the acquisitions clerk was only responsible for the library department. The weekly hours of the acquisition clerk should be expanded from the current part-time status to a full-time status to accommodate this newly-defined role. Upon achieving this goal, the job description for the acquisitions clerk should be rewritten to reflect this change in job responsibilities. The reengineering of this process supports the integration of processes within the Information Resources and Services group.
- The library should establish an electronic reserve system as an option for the students, faculty, and staff of Nichols College. This system would allow some course material that instructors have placed on physical reserve at the library circulation desk to be available for viewing and printing twenty-four hours a day, seven days a week from any computer connected to the Nichols College campus network. Materials placed on reserve by faculty would include course syllabi, lecture notes, exams, exercises, and journal articles. The electronic reserve system should be web-based, and utilize web-based search capability

to identify particular documents by professor's name and/or course number. The system should require that a valid Nichols College user ID and password be provided to access the electronic reserve materials. The electronic reserve materials should be available and presented in PDF format, which requires the installation of Adobe Acrobat Reader for viewing and printing. All computers in the Library should have the Acrobat Reader installed. If access to electronic reserves is requested from outside the library, and the patron does not have the Acrobat Reader installed on the PC or laptop, instructions for downloading (at no charge), installing, and configuring Acrobat Reader program should be provided to the patron. The reengineering of this process supports the integration of processes within the Information Resources and Services group.

- The Information Resources and Services group should publish a monthly electronic newsletter for the benefit of the Nichols College community. This newsletter should include information from within the library, information technology, and telecommunication departments that describe timely issues and upcoming events. The electronic newsletter should also be a medium for providing hints and tips for the effective identification and retrieval of information. The electronic newsletter should be available both in paper format within the library, and in web-based format as a part of the Information Resources and Services website. This goal encourages the integration of all cultures found within the Information Resources and Services group.

- The circulation desk of the library, which is staffed over eighty hours each week, should become the permanent location for the Nichols College telecommunications switchboard. During the last year, the number of student work-study assistants was increased, as the circulation desk began serving as the backup switchboard for the campus. In doing so, the console operator/ receptionist was able to spend time helping other campus administrative departments perform numerous peripheral duties such as admissions and alumni mailings. Once the circulation desk of the library becomes the permanent location for the Nichols College telecommunications switchboard, the role and job description of the console operator/receptionist should be reviewed. The reengineering of this process supports the integration of processes within the Information Resources and Services group.

- The circulation desk of the library, which is staffed over eighty hours each week, should expand its physical reserves to include the temporary loan of audio/ visual and computer equipment utilized by the campus community, The process within the library of loaning printed information and the process within telecommunications of loaning equipment are duplicate processes within the Information Resources and Services group. To facilitate the combining of these duplicate process, a small holding area will need to be cleared behind the circulation desk to accommodate the equipment staged for loan. The reengineering of this process supports the integration of processes within the Information Resources and Services group.

- Based upon the increased demand for information access technology within the library, all desktop computers and printers located within the Information Technology department should be relocated throughout the library. The public access image scanner located within the Information Technology department should also be relocated to the library. Furthermore, an additional image scanner should be placed within the library to satisfy the increased demand for scanner services brought about by the increased interest in webpage design and creation. These physical relocations support the integration of informational processes, and encourage the integration of all cultures found within the Information Resources and Services group.
- The Nichols College web server should be upgraded from a Pentium 133 MHz with 32 MB RAM to a Dual Pentium II 300 MHz with 128 MB RAM. With numerous online interactive forms submitting data to the website, and extensive information being served regarding Nichols College and its curricula, it is imperative that an image and perception of technological excellence be projected. The reengineering of this process supports the integration of processes within the Information Resources and Services group.
- A technology and information retrieval help desk should be created, and located at the circulation desk of the library. This help desk should be "instructional" in nature, and be designed to provide basic technology and information retrieval instruction to the students, faculty, and staff who call for either hardware, software, or information retrieval assistance. This instruction should be limited in scope to the level of assistance that can effectively be provided through a telephone conversation between the caller and the help desk. In addition, a web-based interface should be established so that users have the option to provide the help desk information via the submission of an online form. The creation of an instructional help desk supports the integration of informational processes, and encourages the integration of all cultures found within Information Resources and Services.
- The Information Resources and Services web pages should be expanded to include all of the instructional materials available via telephone at the proposed instructional help desk. By creating a web-based mirror image of the instructional information available at the proposed instructional help desk, an additional avenue for user assistance will be created. The reengineering of this process supports the integration of processes within the Information Resources and Services group.

- A program for the systematic training of student work-study assistants within the library should be initiated. This training should include the general training already provided for library procedures and systems, and then be expanded to include training for the routine software issues that arise from the utilization of library computers. By more fully utilizing the student work force, the Information Resources and Services staff can focus on broader, more proactive issues of support. This type of support training supports the integration of informational processes, and encourages the integration of all cultures found within the Information Resources and Services group.
- It has been a long-standing policy that friends of the library who visit from the surrounding community are allowed to possess a library card, and are consequently allowed to check out printed materials from the library. A guest account should be established that is accessible from all computers within the library. In making this guest account available to those friends of the library who are otherwise not students, faculty, or staff, a broader range of information access will be afforded to all patrons. This account should use the username "guest" and the password "*****". This account should be established such that no e-mail account is available, no physical storage space of college servers is available, and no remote dial-up access is available. Patrons using this guest account would only be able to browse the world wide web from library computers, thereby enhancing their information access.

Goals Beyond the 1999–2000 Academic Year

The long term goals identified for completion during the 2000–2001, 2001–2002, 2002–2003, and 2003–2004 academic years are defined as follows:

- Information Resources and Services should adopt a long term commitment to upgrade all information systems software and hardware as each upgrade becomes available. This must be done in an effort to provide the most efficient and effective computing environment possible to the Nichols College learning community. Furthermore, this action protects the financial and pedagogical investments in technology that Nichols College and its board of trustees have made over recent years.
- Information Resources and Services should identify, select, acquire, manage, and preserve information in all formats in support of its mission and the mission of the College. This on-going process requires that the materials made available support all academic programs, and that continued consortia participation exists in order to perpetuate partnerships for resource and information sharing and staff development opportunities.

- Information Resources and Services should reorganize the Nichols College archival materials into a more efficient and accessible information resource. The archives, currently located in a single room within a room, should be relocated on the first floor of the library building into the area that was formerly utilized as the computer lab within the former Computer Services department. In doing so, the area available for archival materials will quadruple. This relocation will create a more efficient archival operation that fosters an environment where patrons can view archival materials upon demand. Furthermore, climate controls exist in the proposed area that are not available where the archives currently reside. Once reorganized, this area should be made available to the public. Capitalizing upon good community relations, Nichols College should consider availing the physical resources of the archives as a depository for community archival materials.
- Information Resources and Services should develop a formal disaster recovery plan that addresses all of its functional areas of responsibility. The disaster recovery plan should explain in detail what comprise mission critical systems, and how and to whom those systems will be made available during times of disaster. The disaster recovery plan should be written within the framework of other existing disaster recovery plans already adopted in other areas of Nichols College. The disaster recovery plan should be made available for public access through various methods of campus-wide information dissemination, including the use of the Nichols College website.
- The current academic computing environment was first made available in December, 1997. After its third year of operation and service, Information Resources and Services should review and rebuild that academic computing environment during the 2000–2001 academic year. Decisions should be made regarding the continuance of the UNIX-based operating systems or the adoption and migration to an exclusive Windows NT(c) operating environment. In addition, decisions must be made regarding physical hardware – specifically, network storage space for all members of the campus community.
- As a part of its commitment to excellence, Information Resources and Services should monitor customer satisfaction levels on an annual basis. In doing so, a better understanding of the expectations of the campus learning community can be achieved. Furthermore, the results of customer satisfaction surveys will help serve as an indicator of success of the Information Resources and Services efforts to accommodate the needs of students, faculty, and staff.
- Information Resources and Services should continue to provide fast effective access to information resources both within and beyond the physical walls of the library building, as well as beyond the confines what typically is considered normal business hours. This can be accomplished by working to improve

the Information Resources and Services web pages – specifically, in striving to enhance the Library department's web pages to facilitate hyperlinks to electronic information resources, to provide web-based tutorials covering both general and specific research strategies, and to migrate from a Windows(c)-based online collection catalog system to a catalog system that is web-based.

- Information Resources and Services should significantly expand the promotion and marketing of available information resources to its current and potential constituencies. This can be accomplished by providing bibliographic instruction programs, through printed and electronic mediums, and as a result of both formal and ad-hoc group or individual presentations. By increasing a global awareness of available information resources and services, the level of learning can be raised for all who seek to utilize these technology resources.

- Information Resources and Services should, in conjunction with the senior administration, develop a lifecycle funding model that will provide for the automatic replace of desktop computers and printers after a predetermined amount of time. It is imperative that a framework for planned obsolescence be established so that the level of the current technological environment does not deteriorate over time. It is the expectation of each incoming student that the information infrastructure be as technologically current as reasonably possible. It order to meet that expectation, a plan for funding the replacement of computing systems must be defined, developed, budgeted, and implemented.

- Information Resources and Services should strive to provide a physical environment that is conducive to teaching, research, and study. During the 2000–2001 academic year, every effort should be made to complete the current remodeling of the Library department. The cyber-cafe should be completed, complimenting changes already completed. The signage within the Information Resources and Services group should be improved, reflecting to the public the location of services within the newly-integrated group. Furthermore, the new heating and air conditioning system should be expanded to include the lobby and balcony areas of the library.

- Information Resources and Services should develop a formal plan that addresses the continual training of all its staff. The breadth of this training should encompass not only formal instruction received from external sources, but should also include participation in professional seminars and conferences, membership in professional organizations and associations, and the development of an in-house knowledge reservoir comprised of current technical references found in the form of books, videos, and computer-based tutorials (CBT). By developing such a plan, staff will be able to discover, develop, employ, and maintain the most effective and up-to-date information technologies for the efficient and effective delivery of informational instruction, services, and resources.

It is believed that through the accomplishment of both the short term and long term goals defined in this strategic plan, a stronger and more positive influence will come to bear upon the Nichols College learning community. Effective strategies for information access will be developed and maintained. A competent methodology for the research of printed and electronic information will be developed and promoted. This continual development and enhancement of campus information resources and services through the accomplishment of these well-defined goals will foster an environment that solidifies the management of knowledge across all lines of content. Improving information resources and services will serve in concert with academic curriculum to develop and enhance student information literacy skills. Once these information literacy skills are positively affected, the information competency of every member of the Nichols College learning community will be elevated.

Annual Assessment and Review

In order for this strategic plan to remain a viable and effective document during the course of its proposed span, Information Resources and Services will, in conjunction with the senior administration, perform an annual assessment and review of the strategic plan. The purpose of this annual exercise will be twofold. First, an annual assessment of the plan's short term goals will be conducted. This assessment will identify those goals accomplished, and those goals remaining to be achieved. Second, an in-depth review of unaccomplished goals will be conducted.

As a result of this review, the short term goals not accomplished will be reviewed to determine if they need to continue to be short term goals, if they need to be revised in any way, or if they need to either be eliminated or redefined as long term goals. Furthermore, the remaining long term goals will be reviewed so that a new group of short term goals for the upcoming year can be identified. The end result of the annual assessment and review will be that as the learning environment at Nichols College continues to grow and evolve, new goals – both short term and long term – will to be identified and added to the strategic plan.

Through annual assessment and review, this strategic plan will become a viable document that is redefined and perpetuated each year. This methodology will involve the senior administration in the strategic planning process for information technology.

Implementation of the Strategic Plan

The chief information officer is responsible for the implementation of this strategic plan for the Information Resources and Services group at Nichols College. As outlined in this strategic plan, a timeline for implementation will commence with the 1999–2000 academic year. Based upon the successful outcome of annual assessment and review, this strategic plan will perpetuate itself each year, based upon new sets of both short term and long term goals being defined and adopted. This strategic plan is available on the Nichols College website for public review, and as a result of annual assessment and review, will immediately reflect all changes in strategic direction as deemed appropriate.

Conclusion

In order for this strategic plan to be effective, it is imperative that all campus constituencies provide continual review and support of the goals we have now set for ourselves and for our customers. The thoughts and concerns of many have come together during the creation of this strategic plan. The values built into each of these goals reflect the innovation, creativity, service, commitment, knowledge, appreciation's of diversity and ownership, and mission-oriented teamwork that Information Resources and Services accepts as the framework for its purpose within the academic community.

The successful attainment of our strategic plan goals will inspire a technologically superior "information commons" that will undoubtedly become a central meeting place for all who are or will someday become a part of the learning village known as Nichols College.

REFERENCES

Aebersold, D., & Haaland, G. A. (1994). *Strategic information resources: A new organization* [Online]. Available: http://www.educause.edu/ir/library/text/cnc9418.txt [1998, November 8].

Aken, R., & Molinaro, M. (1995). *What I really want from the world wide web is . . . : How do we establish access to ever expanding resources?* [Online]. Available: http://www.educause.edu/ir/library/word/cnc9558.word [1998, November 22].

American Library Association (1989, January). *American Library Association presidential committee on information literacy.* Washington, DC: Author.

Anderson, D., & Gesin, J. (1997). *The evolving roles of information professionals in the digital age* [Online]. Available: http://www.cause.org/information-resources/ir-library/html/ cnc9754/cnc9754.html [1998, November 18].

Anglin, D. L. (1996). *Development of a guide for strategic planning for Assemblies of God education institutions and ministries in Asia Pacific*. Unpublished doctoral dissertation, Nova Southeastern University, Fort Lauderdale, FL.

Arzt, N., Beck, R., Gordon, J., & May, L. (1993). *Architecture and re-engineering: Partnership for change at the University of Pennsylvania* [Online]. Available: http://www.educause.edu/ir/library/pdf/cnc9313.pdf [1998, October 9].

Barone, C. A. (1996). Leading through influence [Online]. Available: http://www.educause.edu/ir/library/pdf/pub3015.pdf [1998, October 2].

Bass, B. M. (1992). *Bass & Stogill's handbook of leadership: Theory, research, and managerial applications* (3rd ed.). New York, NY: The Free Press.

Battin, P. (1996). *Making it happen: Leadership in a transformational age* [Online]. Available: http://www.educause.edu /ir/library/pdf/pub3015.pdf [1998, October 2].

Battin, P., & Hawkins, B. L. (1997). The changing role of the information resources professional: A dialogue. *Cause/Effect, 20*(1), 22–30.

Beltrametti, M. (1993). *Integrating computing resources: A shared distributed architecture for academics and administrators* [Online]. Available: http://www.educause.edu/ir/library/word/cnc9302.word [1998, October 26].

Bernbom, G., & Hess, C. (1994). *INforum: A library/IT collaboration in professional development at Indiana University* [Online]. Available: http://www.educause.edu/ir/library/word/cem9434.word [1998, October 16].

Bleed, R., & McClure, P. A. (1995). *What information resources managers need to understand about the higher education enterprise* [Online]. Available: http://www.educause.edu/ir/library/pdf/cem954a.pdf [1998, October 20].

Blunt, C. R. (1993). *Some college/university roles in the transition to an information age society* [Online]. Available: http://www.educause.edu/ir/library/word/cnc9336.word [1998, November 6].

Bortz, W. M. (1993). *Implementing a culture of change: The five year transformation of The George Washington University* [Online]. Available: http://www.educause.edu/ir/library/word/cnc9309.word [1998, October 4].

Branson, R. K. (1990, April). Issues in the design of schooling: Changing the paradigm. *Educational Technology, 30,* 7–10.

Brentrup, R., Fark, R., Jargo, L., & LaMarca, M. (1997). *Managing a collection of electronic information resources* [Online]. Available: http://www.educause.edu/ir/library/html/cnc9753/cnc9753.html [1998, October 23].

Brodie, M., & McLean, N. (1995). *Process reengineering in academic libraries: Shifting to client-centered resource provision* [Online]. Available: http://www.educause.edu/ir/library/word/cnc9528.word [1998, November 23].

Brown, J. K. (1999). Libraries face the 21st century. *Converge, 2*(2), 40–41.

Bunker, G., & Horgan, B. (1994). *Customer-centered collaboration: Libraries and IT* [Online]. Available: http://www.educause.edu/ir/library /word/cnc9409.word [1998, October 29].

Carter, G. F., Lundy, H. W., & Penn, J. D. (1993). *Where do we go from here: Summative assessment of a five-year strategic plan for linking and integrating information resources* [Online]. Available: http://www.educause.edu/ir/library/word/cnc9333.word [1998, November 5].

Cisneros, L., Hunt, D., & McCollam, D. (1997). *Information architecture: Bringing the university's information inventory under control* [Online]. Available: http://www.cause.org/information-resources/ir-library/html/cnc9759/cnc9759.html [1998, November 20].

Clayton, S. R. (1997). *Is the end-user now the head of collection development?: Distributed computing issues for academic libraries* [Online]. Available: http://www.cause.org/information-resources/ir-library/html/cnc9749/cnc9749.html [1998, November 17].

Conley, D. B., Detweiler, R. A., Falduto, E. F., & Golden, R. M. (1996). *IT's in the plan: Integrating institutional and IT planning (lessons for faculty advocates, planners, technologists, CIOs, and presidents)* [Online]. Available: http://www.educause. edu/ir/library/pdf/cnc9636.pdf [1998, November 11].

Conrad, L., Rome, J., & Wasileski, J. (1993). *Will the last central IT person please turn off the lights?* [Online]. Available: http://www.educause.edu/ir/library/word/cnc9218.word [1998, October 12].

Creth, S. D. (1994). Organization design: New paths for collaboration. In: S. D. Creth & A. G. Lipow (Eds), *Computing and Library Professionals: Building Partnerships* (pp. 11–22). San Carlos, CA: Library Solutions Press.

DeMauro, K. (1997). *Improve morale and reduce stress: Communicate!* [Online]. Available: http://www.cause.org/information-resources/ir-library/html/cnc9711.html [1998, November 13].

DeNoia, L. A. (1993). *The art and politics of re-engineering under crisis conditions* [Online]. Available: http://www.educause. edu/ir/library/word/cnc9315.word [1998, November 4].

DeSantis, D. J., & Laudato, N. C. (1995). *Reshaping the enterprise through an information architecture and process reengineering* [Online]. Available: http://www.educause.edu/ir/library/pdf/cem9546.pdf [1998, October 19].

Detloff, J., Ecelbarger, D., & Wilburn, K. (1996). *An integrated information environment: A technology architecture for the next century* [Online]. Available: http://www.educause.edu/ir/library/pdf/cnc9631.pdf [1998, October 11].

Dolence, M. G., Douglas, J. V., & Penrod, J. I. (1990). *The chief information officer in higher education* (CAU.S.E Professional Paper 4). Boulder, CO: CAU.S.E.

Duwe, J., Luker, M., & Pinkerton, T. (1995). *Restructuring a large IT organization: Theory, model, process, and initial results* [Online]. Available: http://www.educause.edu/ir/library/pdf/cem9525.pdf [1998, October 28].

Eaton, R. B., & Schuler, R. C. (1994). *Revolution in the information systems shop: Reengineering the IS workplace* [Online]. Available: http://www.educause.edu/ir/library/word/cem9416.word [1998, October 22].

Ernst, D. J., Katz, R. N., & Sack, J. R. (1994). *Organizational and technological strategies for higher education in the information age* (CAU.S.E Professional Paper 13). Boulder, CO: CAU.S.E.

Eustis, J., & McMillan, G. (1997). *Technology initiatives and organizational change: Higher education in a networked world* [Online]. Available: http://www.cause.org/information-resources/ ir-library/html/cnc9756/cnc9756.html [1998, November 19].

Fleit, L. H. (1994). *Self-assessment for campus information technology services* (CAU.S.E Professional Paper 12). Boulder, CO: CAU.S.E.

Foley, T. J. (1997). *Combining libraries, computing, and telecommunications: A work in progress* [Online]. Available: http://www.lehigh.edu/tjf0/public/www-data/acm97/acm97.htm [1998, June 4].

Fox, K. C. (1998). *Information technology in higher education: Evolving learning environments* [Online]. Available: http://www.educause.edu/ir/library/pdf/cmr9823.pdf [1998, November 2].

Franklin, S. D. (1994). Libraries and academic computing centers: Forging new relationships in a networked information environment. In: S. D. Creth & A. G. Lipow (Eds), *Computing and Library Professionals: Building Partnerships* (pp. 23–30). San Carlos, CA: Library Solutions Press.

Gabel, M. R., Hurt, C. S., O'Connor, J. S., & Sevon, W. W. (1995). *Emerging environments and professionals: The university center as catalyst for GMU's vision* [Online]. Available: http://www.educause.edu/ir/library/pdf/cnc9532.pdf [1998, October 25].

Gleason, B. W. (1996). Managing Ideas [Online]. Available: http://www.educause.edu/ir/library/ pdf/pub3015.pdf [1998, October 2].

Glenn, T. J., & Mechley, V. P. (1993). *Empowering the user* [Online]. Available: http://www. educause.edu/ir/library/word/cnc9310.word [1998, October 6].

Goetsch, L., & Kaufman, P. T. (1997). *Readin', writin', arithmetic, and information competency: Adding a basic skills component to a university's curriculum* [Online]. Available: http://www.educause.edu/ir/library/html/cnc9750/cnc9750.html [1998, December 6].

Goguen, B. J., & Hogue, W. F. (1997). *Reengineering customer assistance: The good, the bad, and the ugly* [Online]. Available: http://www.cause.org/information-resources/ir-library/html/ cnc9748/cnc9748.html [1998, November 16].

Gregorian, V., Hawkins, B., & Taylor, M. (1993). *What president's need to know about the integration of information technologies on campus* [Online]. Available: http://www.educause. edu/ir/library/text/hei1010.txt [1998, October 15].

Groff, W. H. (1994). *Toward the 21st century: Preparing proactive visionary transformational leaders for building learning communities.* Fort Lauderdale, FL: Nova Southeastern University. (ERIC Document Reproduction Service No. 372 239)

Hardesty, L. (1998). Computer center-library relations at smaller institutions: A look from both sides. *Cause/Effect, 21*(1), 35–41.

Hardesty, L. (1999). *First depress clutch, then shift paradigm* [Online]. Available: http://library.willamette.edu/home/   publications/movtyp/fall1998/hardesty.htm   [1999, January 5].

Hawkins, B. L. (1996). *Leadership in a service environment* [Online]. Available: http://www. educause.edu/ir/library/pdf/ pub3015.pdf [1998, October 2].

Heterick, R. C. (1996). *Maybe Adam Smith had it right* [Online]. Available: http://www. educause.edu/ir/library/pdf/ pub3015.pdf [1998, October 2].

Higginbotham, B. B. (1997). *Marriage, Brooklyn style: How the library, the faculty, and administrative computing deliver Brooklyn College's academic computing program* [Online]. Available: http://www.cause.org/information-resources/ir-library/html/cnc9731/cnc9731.html   [1998, November 15].

Hirshon, A. (1998). *Integrating computing and library services: An administrative planning and implementation guide for information resources* (CAU.S.E Professional Paper 18). Boulder, CO: CAU.S.E.

Hirshon, A. (1997). *Merging libraries and computing: Institutional decision-making and implementation* [Online]. Available: http://www.cause.org/information-resources/ir-library/html/ cnc9726/cnc9726.html [1998, November 14].

Hirshon, A. (1996). *Strategic planning and restructuring of Lehigh University information resources* [Online]. Available: http://www.lehigh.edu/~inluir/irdocs/reorgtxt.htm [1998, September 27].

Hofstetter, L. M., & Mullin, M. E. (1993). *Doing more with less: A pragmatic approach to getting work done* [Online]. Available: http://www.educause.edu/ir/library/word/cnc9316.word [1998, October 5].

John, N. R. (1996). *Putting content onto the Internet: The library's role as creator of electronic information* [Online]. Available: http://www.uic.edu/ñrj/fm.html [1998, November 24].

Katz, R. N., & West, R. P. (1993). *Sustaining Excellence in the 21st century: A vision and strategies for college and university administration* (CAU.S.E Professional Paper 8). Boulder, CO: CAUSE.

Kiesler, S. (1994). Working together apart. In: S. D. Creth & A. G. Lipow (Eds), *Computing and Library Professionals: Building Partnerships* (pp. 1–10). San Carlos, CA: Library Solutions Press.

Kim, Y. G. (1998). *A strategic plan for the integration of technology into the elementary teacher education program at Inchon National University of Education.* Unpublished doctoral dissertation, Nova Southeastern University, Fort Lauderdale, FL.

LeDuc, A. L. (1996a). *Organizational leadership: characteristics of success and failure* [Online]. Available: http://www.educause.edu/ ir/library/pdf/pub3015.pdf [1998, October 2].

LeDuc, A. L. (1996b). *Personnel development: Key to organizational strength* [Online]. Available: http://www.educause.edu/ir/library/pdf/cnc9645.pdf [1998, November 12].

Lindauer, B. G., Oberman, C., & Wilson, B. (1998). Integrating information literacy into the curriculum: How is your library measuring up? *College & Research Libraries News, 59*(5), 347–352.

Lowry, A. (1993). *The information arcade* [Online]. Available: http://www.educause.edu/ir/library/ word/cnc9346.word [1998, November 7].

MacKnight, C. B. (1995). *Managing technological change in academe* [Online]. Available: http://www.educause.edu/ir/library/word/cem9517.word [1998, October 14].

Mancuso, R., & Stahl, B. (1996). *Implementing a combined computing services and library help desk* [Online]. Available: http://www.educause.edu/ir/library/pdf/cnc9617.pdf [1998, November 10].

Massey, M. G., & Stedman, D. W. (1995). *Emotional climate in the information technology organization: Crisis or crossroads?* [Online]. Available: http://www.educause.edu/ir/library/pdf/ cem9543.pdf [1998, October 21].

Massy, W. F., & Zemsky, R. (1995). *Using information technology to enhance academic productivity* [Online]. Available: http://www.educause.edu/nlii/keydocs/massy.html [1998, October 31].

Mayer, R. (1992). Cognition and instruction: Their historic meeting within educational psychology. *Journal of Educational Psychology, 84*(4), 405–412.

McClintock, M. (1997). *The role of the information systems professional in small colleges and universities* [Online]. Available: http://www.cause.org/information-resources/ir-library/ html/cnc9764/cnc9764.html [1998, November 21].

McClure, P. A., Smith, J. W., Stager, S. F., & Williams, J. G. (1993). *Assessing the effectiveness of information technology* [Online]. Available: http://www.educause.edu/ir/library/word/ cnc9324.word [1998, October 10].

McClure, P. A., Sitko, T. D., & Smith, J. W. (1997). *The crisis in information technology support: Has our model reached its limit?* (CAU.S.E Professional Paper 16). Boulder, CO: CAU.S.E.

Metz, T. (1999). *Academic libraries and computing centers: The case for collaboration* [Online]. Available: http://library.willamette.edu/home/publications/movtyp/spring99/Metz.html [1999, January 6].

Molinaro, M. (1997). *Changing our focus: Creating a user-centered organization* [Online]. Available: http://www.cause.edu/information-resources/ir-library/html/cnc9742/ cnc9742.html [1998, October 8].

Morgan, E. L. (1998). Communication is the key to our success. *Computers in Libraries, 18*(9), 28–30.

Nasseh, B. (1998). *Key planning issues in technology-based education: Elements of effective strategical planning for technology in education* [Online]. Available: http://www.educause. edu/ir/library/pdf/cmr9814.pdf [1998, November 1].

Overlock, T. (1995). *Development of a multiyear plan for the integration of multimedia into the learning environment at Northern Maine Technical College.* Unpublished doctoral dissertation, Nova Southeastern University, Fort Lauderdale, FL.

Pecka, K. (1993). *User-driven training: A strategy for support* [Online]. Available: http://www. educause.edu/ir/ library/word/cnc9311.word [1998, November 3].

Penrod, J. I. (1996). *Critical process redesign (CPR) to achieve an IT learning organization* [Online]. Available: http://www.educause.edu/ir/library/word/cnc9647.word [1998, October 24].

Pflueger, K. E. (1995). *Collaborating for the more effective integration and use of technology: Rethinking user services* [Online]. Available: http://www.educause.edu/ir/library/word/ cnc9528.word [1998, October 18].

Rapple, B. A. (1997). *The electronic library: New roles for librarians* [Online]. Available: http://www.cause.org/information-resources/ir-library/html/cem971a.html [1998, October 1].

Ringle, M., & Updegrove, D. (1997). *Is strategic planning for technology an oxymoron?* [Online]. Available: http://www.cause.org/information-resources/ir-library/html/cnc9758/cnc9758.html [1998, October 27].

Roberts, L. H. (1996). *The network as a strategic asset: The path to learning productivity* [Online]. Available: http://www.educause.edu/nlii/articles/roberts.html [1998, October 30].

Rocheleau, B. (1996). *Structures, plans, and policies: Do they make a difference?* [Online]. Available: http://www.cause. org/information-resources/ir-library/html/cem9637.html [1998, October 13].

Sharrow, M. J. (1995). *Library and IT collaboration projects: Nine challenges* [Online]. Available: http://www.educause.edu/ir/library/pdf/cem9540.pdf [1998, October 17].

Silverman, S., & Silverman, G. (1999). The Educational Enterprise Zone: Where knowledge comes from! *Technological Horizons in Education Journal, 26*(7), 56–57.

Smith, S. L., & West, S. M. (1995). *Library and computing merger: Clash of titans or golden opportunity* [Online]. Available: http://www.educause.edu/ir/library/word/cnc9564.word [1998, November 9].

Sonntag, G. (1997). Our raison d'àtre: Teaching information competencies. *College & Research Libraries News, 58*(11), 770–771.

Stahl, B. (1995). *Trends and challenges for academic libraries and information services* [Online]. Available: http://www.educause.edu/ir/ library/word/cem951a.word [1998, October 7].

West, T. W. (1996). *More lessons from the CIO trail* [Online]. Available: http://www.educause.edu/ ir/library/pdf/ pub3015.pdf [1998, October 2].

Young, A. P. (1994). *Information technology and libraries: A virtual convergence* [Online]. Available: http://www.educause.edu/ ir/library/word/cem9431.word [1999, January 16].

ADDITIONAL READINGS

Allen, B. M., & Gosling, W. A. (1997). Facing change and challenge through collaborative action: The CIC libraries' experience. In: C. A. Schwartz (Ed.), *Restructuring academic libraries: Organizational development in the wake of technological change* (pp. 121–138). Chicago, IL: American Library Association.

Baker, G. S. (1994). The Knoxville library/computer center partnership. In: S. D. Creth & A. G. Lipow (Eds), *Computing and Library Professionals: Building Partnerships* (pp. 51–53). San Carlos, CA: Library Solutions Press.

Bates, A. W. (1997). Restructuring the university for technological change [Online]. Available: http://bates.cstudies.ubc.ca/carnegie/ carnegie.html [1999, January 3].

Berger, M. A. (1999). Tech support people: We're not that bad! *Computers in Libraries, 19*(1), 8.

Bianchi, J., Pflueger, K. E., & Thompson, C. (1998). *The emergence of convergence: An interactive reflection on the future of information services and information technology organizations* [Online]. Available: http://www.usc.edu/isd/projects/emergeconverge/clupresentgray.pdf [1999, January 4].

Billings, H. (1998). Libraries, language, and change: Defining the information present. *College & Research Libraries, 59*(3), 212–218.

Birdsall, D. G. (1997). Strategic planning in academic libraries: A political perspective. In: C. A. Schwartz (Ed.), *Restructuring Academic Libraries: Organizational Development in the Wake of Technological Change* (pp. 253–261). Chicago, IL: American Library Association.

Blair, J. (1992). The library in the information revolution. *Library Administration & Management, 6*(2), 71–76.

Blustain, H., Goldstein, P., & Lozier, G. (1999). Assessing the new competitive landscape. In: R. N. Katz (Ed.), *Dancing With the Devil: Information Technology and the New Competition in Higher Education* (pp. 51–71). San Francisco, CA: Jossey-Bass Publishers.

Bly, O., & Wiggins, R. (1994). A legacy of partnership: The University of Akron's libraries and computer center. In: S. D. Creth & A. G. Lipow (Eds), *Computing and Library Professionals: Building Partnerships* (pp. 55–58). San Carlos, CA: Library Solutions Press.

Bryson, J. (1997). *Managing information services: An integrated approach.* Brookfield, VT: Gower House.

Butler, M. A. & DeLong, S. E. (1997). Planning information systems at the University of Albany: False starts, promising collaborations, evolving opportunities. In: C. A. Schwartz (Ed.), *Restructuring Academic Libraries: Organizational Development in the Wake of Technological Change* (pp. 81–95). Chicago, IL: American Library Association.

Chatterji, M. K., Davis, H. S., Martin, R. G., & Toulson, R. (1994). Consolidation, centralization, connections, and cooperation at Indiana State University. In: S. D. Creth & A. G. Lipow (Eds), *Computing and Library Professionals: Building Partnerships* (pp. 65–67). San Carlos, CA: Library Solutions Press.

Cook, B. (1995). *Managing for effective delivery of information services in the 1990s: The integration of computing, educational and library services* [Online]. Available: http://online.anu.edu.au/CNASI/pubs/OnDisc95/docs/ONL05.html [1998, December 2].

Corbin, J. (1992). Technical services for the electronic library. *Library Administration & Management, 6*(2), 86–90.

Corbus, L. (1999). Taking charge of micromanagers. *American Libraries, 30*(2), 26–28.

Creth, S. D. (1993). Creating a virtual information organization: Collaborative relationships between libraries and computing centers. *Journal of Library Administration, 19*(3/4), 111–132.

Davis, F. C. (1995). *Finally! Online access to databases: A tale of woe, wonder, wizards and witches* [Online]. Available: http://www.dowling.edu/library/papers/CIL/FINALLY/Finally.htm [1999, January 17].

Davis, F. C., & Gotsch, J. R. (1999). *The essential information specialist* [Online]. Available: http://www.dowling. edu/library/papers/CIL/TURKO1/Turkeys.htm [1999, January 16].

Davis-Millis, N., & Owens, T. (1997). Two cultures: A social history of the distributed library initiative at MIT. In: C. A. Schwartz (Ed.), *Restructuring Academic Libraries: Organizational Development in the Wake of Technological Change* (pp. 96–107). Chicago, IL: American Library Association.

Dockstader, J. (1999). Teachers of the 21st century know the what, why, and how of technology integration. *Technological Horizons in Education Journal, 26*(6), 73–74.

Doughty, M. C., Nielsen, B., & Steffen, S. S. (1995). *Computing center/library cooperation in the development of a major university service: Northwestern's electronic reserve system* [Online]. Available: http://www.educause.edu/ir/library/ word/cnc9561.word [1999, January 5].

Dougherty, R. M., & McClure, L. (1997). Repositioning campus information units for the era of digital libraries. In: C. A. Schwartz (Ed.), *Restructuring Academic Libraries: Organizational*

Development in the Wake of Technological Change (pp. 67–80). Chicago, IL: American Library Association.

Duderstadt, J. J. (1999). Can colleges and universities survive in the information age? In: R. N. Katz (Ed.), *Dancing with the Devil: Information Technology and the New Competition in Higher Education* (pp. 1–25). San Francisco, CA: Jossey-Bass Publishers.

Euster, J. R., Kaufman, J., Paquette, J., & Soete, G. (1997). Reorganizing for a changing information world. *Library Administration & Management, 11*(2), 103–114.

Farrington, G. C. (1999). The new technologies and the future of residential undergraduate education. In: R. N. Katz (Ed.), *Dancing with the Devil: Information Technology and the New Competition in Higher Education* (pp. 73–94). San Francisco, CA: Jossey-Bass Publishers.

Fishel, T., & Shultz, P. (1994). Areas of cooperation between the library and computing. In: S. D. Creth & A. G. Lipow (Eds), *Computing and Library Professionals: Building Partnerships* (pp. 59–60). San Carlos, CA: Library Solutions Press.

Flowers, K., & Martin, A. (1994). *Enhancing user services through collaboration at Rice University* [Online]. Available: http://www.educause.edu/ir/library/word/cem9435.word [1999, January 6].

Foley, T. J. (1998). *The metamorphosis of libraries, computing, and telecommunications into a cohesive whole* [Online]. Available: http://www.educause.edu/ir/library/pdf/cmr9837.pdf [1999, January 2].

Forys, M., & Simmons-Welburn, J. (1994). Library explorer: Information skills, strategies, and sources. In: S. D. Creth & A. G. Lipow (Eds), *Computing and Library Professionals: Building Partnerships* (pp. 73–75). San Carlos, CA: Library Solutions Press.

Giesecke, J., & Walter, K. (1997). Adapting organizational structures in technical services to new technologies: A case study of the University of Nebraska-Lincoln libraries. In: C. A. Schwartz (Ed.), *Restructuring Academic Libraries: Organizational Development in the Wake of Technological Change* (pp. 192–199). Chicago, IL: American Library Association.

Gleason, B. W. (1993). *A valuable lesson: Trust in people, the rest is easy* [Online]. Available: http://www.educause.edu/ir/ library/word/cem9312.word [1999, January 7].

Gleason, M., & Ostlund, J. (1994). Documents to the people: A collaborative approach. In: S. D. Creth & A. G. Lipow (Eds), *Computing and Library Professionals: Building Partnerships* (pp. 71–72). San Carlos, CA: Library Solutions Press.

Goodson, C. (1997). Putting the service back in library service. *College & Research Libraries News, 58*(3), 186–187.

Grass, C., & Hughes, J. (1994). Building a team-based, integrated information services organization. In: S. D. Creth & A. G. Lipow (Eds), *Computing and Library Professionals: Building Partnerships* (pp. 61–63). San Carlos, CA: Library Solutions Press.

Grassian, E. (1997). *Information literacy competencies: Selected items & efforts* [Online]. Available: http://www.ala.org/acrl/nili/ competen.html [1999, February 13].

Graves, W. H. (1999). Developing and using technology as a strategic asset. In: R. N. Katz (Ed.), *Dancing with the Devil: Information Technology and the New Competition in Higher Education* (pp. 95–118). San Francisco, CA: Jossey-Bass Publishers.

Groff, W. H. (1996a). *Creating and sustaining learning communities*. Fort Lauderdale, FL: Nova Southeastern University. (ERIC Document Reproduction Service No. 389 890)

Groff, W. H. (1996b). *Creating and sustaining learning communities in the digital era*. Fort Lauderdale, FL: Nova Southeastern University. (ERIC Document Reproduction Service No. 396 188)

Hatfield, R. L. (1998). *Improving team performance with facilitation* [Online]. Available: http://www.educause.edu/ir/ library/pdf/cmr9821.pdf [1999, January 9].

Hersberger, R. M. (1997). Leadership and management of technological innovation in academic libraries. *Library Administration & Management, 11*(1), 26–29.

Highfill, W. C., & Medina, S. O. (1997). Shaping consensus: Structured cooperation in the network of Alabama academic libraries. In: C. A. Schwartz (Ed.), *Restructuring academic libraries: Organizational Development in the Wake of Technological Change* (pp. 139–159). Chicago, IL: American Library Association.

Katz, R. N. (1999). Competitive strategies for higher education in the information age. In: R. N. Katz (Ed.), *Dancing with the Devil: Information Technology and the New Competition in Higher Education* (pp. 27–49). San Francisco, CA: Jossey-Bass Publishers.

Kent, C. M. (1997). Rethinking public services at Harvard College library: A case study of co-ordinated decentralization. In: C. A. Schwartz (Ed.), *Restructuring Academic Libraries: Organizational Development in the Wake of Technological Change* (pp. 180–191). Chicago, IL: American Library Association.

Kesner, R. M. (1997). *Developing an information technology support model for higher educa-tion* [Online]. Available: http://www.educause.edu/ir/library/html/cem9725.html [1999, January 8].

Kirkpatrick, T. E. (1998). The training of academic library staff on information technology within the libraries of the Minnesota state colleges and universities system. *College & Research Libraries, 59*(1), 51–59.

Kohl, D. F. (1997). Farewell to all that . . . Transforming collection development to fit the virtual library context: The OhioLINK experience. In: C. A. Schwartz (Ed.), *Restructuring Academic Libraries: Organizational Development in the Wake of Technological Change* (pp. 108–120). Chicago, IL: American Library Association.

Lakos, A. (1998). Performance measurement in libraries and information services. *College & Research Libraries News, 59*(4), 250–251.

Lewis, D. W. (1997). Change and transition in public services. In: C. A. Schwartz (Ed.), *Restructuring Academic Libraries: Organizational Development in the Wake of Technological Change* (pp. 31–53). Chicago, IL: American Library Association.

Lindauer, B. G. (1998). Defining and measuring the library's impact on campuswide outcomes. *College & Research Libraries, 59*(6), 546–563.

Lynch, M. J. (1996a). Electronic services: Who's doing what? *College & Research Libraries News, 57*(10), 661–663.

Lynch, M. J. (1996b). How wired are we? New data on library technology. *College & Research Libraries News, 57*(2), 97–100.

Maloney, S. (1997). Computers in libraries '97: Looking for quality. *College & Research Libraries News, 58*(6), 408–409.

Martin, M. J. (1992). Academic libraries and computing centers: Opportunities for leadership. *Library Administration & Management, 6*(2), 77–81.

Martin, R. R. (1997). Restructuring the University of Vermont libraries: Challenges, opportunities, and change. In: C. A. Schwartz (Ed.), *Restructuring Academic Libraries: Organizational Development in the Wake of Technological Change* (pp. 168–179). Chicago, IL: American Library Association.

Miller, G. (1996). *Customer service & innovation in libraries*. Fort Atkinson, WI: Highsmith Press.

Mullins, J. L. (1994). An opportunity: Cooperation between the library and computer services. In: S. D. Creth & A. G. Lipow (Eds), *Computing and Library Professionals: Building Partnerships* (pp. 69–70). San Carlos, CA: Library Solutions Press.

Nelson, M. R. (1999). *We have the information you want, but getting it will cost you: Being held hostage by information overload* [Online]. Available: http://info.acm.org/crossroads/Xrds1-1/mnelson.html [1999, February 15].

Nyce, J. M., & Stahlke, H. (1998). *Common sense, traditional structures and higher education: Reengineering as fundamental inquiry* [Online]. Available: http://www.educause.edu/ir/library/pdf/cmr9807.pdf [1999, January 10].

Oblinger, D. G., & Verville, A. (1999). Information technology as a change agent. *Educom Review, 34*(1), 46–55.

Ohr, D. M., & Sonntag, G. (1996). The development of a lower-division, general education, course-integrated information literacy program. *College & Research Libraries, 57*(4), 331–338.

Olcott, L. M. (1998). *Dealing with change: A user's perspective* [Online]. Available: http://www.educause.edu/ir/library/pdf/cmr9810.pdf [1999, January 11].

Osburn, C. B. (1997). One purpose: The research university and its library. In: C. A. Schwartz (Ed.), *Restructuring Academic Libraries: Organizational Development in the Wake of Technological Change* (pp. 238–252). Chicago, IL: American Library Association.

Pitkin, G. M. (1993). *Leadership and the changing role of the chief information officer in higher education* [Online]. Available: http://www.educause.edu/ir/library/text/cnc9305.txt [1999, January 7].

Rader, H. B. (1998). Information literacy: The professional issue. *College & Research Libraries News, 59*(3), 171–172.

Rickard, W. (1999). Technology, higher education, and the changing nature of resistance. *Educom Review, 34*(1), 42–45.

Roark, D. B. (1997). Directed technological change in the Florida Community College system. In: C. A. Schwartz (Ed.), *Restructuring Academic Libraries: Organizational Development in the Wake of Technological Change* (pp. 160–167). Chicago, IL: American Library Association.

Roecker, F. (1996). *Beyond the internet: Integrating print, electronic and web resources for users on the WWW gateway to information* [Online]. Available: http://www.educause.edu/ir/library/word/cnc9655.word [1999, January 12].

Scherrei, R. A. (1997). Caught in the crossfire: Organizational change and career displacement in the University of California libraries. In: C. A. Schwartz (Ed.), *Restructuring Academic Libraries: Organizational Development in the Wake of Technological Change* (pp. 231–237). Chicago, IL: American Library Association.

Schwartz, C. A. (1997). Restructuring academic libraries: Adjusting to technological change. In: C. A. Schwartz (Ed.), *Restructuring Academic Libraries: Organizational Development in the Wake of Technological Change* (pp. 1–30). Chicago, IL: American Library Association.

Seiden, P. (1997). Restructuring liberal arts college libraries: Seven organizational strategies. In: C. A. Schwartz (Ed.), *Restructuring Academic Libraries: Organizational Development in the Wake of Technological Change* (pp. 213–230). Chicago, IL: American Library Association.

Shepherd, B. A., Snyder, C. A., & Starratt, J. (1995). *Library as perpetual partner: Providing information access and technology support* [Online]. Available: http://www.educause.edu/ir/library/word/cnc9559.word [1999, January 13].

Smith, A. (1999). *Information literacy* [Online]. Available: http://inst.augie.edu/āsmith/infolit.html [1999, February 14].

Soine, R. (1998). Technology in the trenches: Improving the quality of instruction. *Technological Horizons in Education Journal, 26*(5), 51–52.

St. Clair, G. (1997). Benchmarking and restructuring at Penn State libraries. In: C. A. Schwartz (Ed.), *Restructuring Academic Libraries: Organizational Development in the Wake of Technological Change* (pp. 200–212). Chicago, IL: American Library Association.

Starratt, J. (1998). Librarians as university technology leaders. *Library Administration & Management, 12*(4), 188–194.

Sweeney, R. (1997). Leadership skills in the reengineering library. *Library Administration & Management, 11*(1), 30–41.

Tanji, L. (1994). UCI libraries internet information access via world-wide web. In: S. D. Creth & A. G. Lipow (Eds), *Computing and Library Professionals: Building Partnerships* (pp. 77–79). San Carlos, CA: Library Solutions Press.

Walczak, F. R. (1998). Technology integration: An age old problem. *Converge, 1*(4), 80–81.

West, R. P. (1994a). *CNI: A successful merger of content and distribution* [Online]. Available: http://www.educause.edu/ir/library/word/cem9432.word [1999, January 14].

West, R. P. (1994b). *Library budgets: Re-orienting where we spend our money* [Online]. Available: http://www.educause.edu/ir/library/word/cem9421.word [1999, January 15].

White, H. S. (1997). Dangerous misconceptions about organizational development of virtual libraries. In: C. A. Schwartz (Ed.), *Restructuring Academic Libraries: Organizational Development in the Wake of Technological Change* (pp. 54–66). Chicago, IL: American Library Association.

A RECOMMENDED METHODOLOGY FOR DETERMINING THE DISPARITY BETWEEN WOMEN'S SALARY LEVELS AND THOSE OF MEN IN THE LIBRARIAN PROFESSORATE IN AN ACADEMIC LIBRARY SETTING

Elizabeth A. Titus

INTRODUCTION

In 1977, the American Association of University Professors (AAUP) published a *Higher Education Salary Kit* by Elizabeth L. Scott on the results and suggested costs of several suggested methods for addressing salary inequities of women and minority faculty, which included a "recommended method for flagging women and minority persons for whom there is apparent salary inequity" (Scott, E. L., 1977, title page). The objective of this study was not just to show differences between the average salaries of male and female faculty within an institution or group ("Average faculty", 1996; "Annual report of the economic status", 1996). Scott argued that salary studies which showed differences in the average salaries between male and female faculty did not establish and measure gender-based salary discrimination in faculty salaries.

Advances in Library Administration and Organization, Volume 18, pages 123–173.
Copyright © 2001 by Elsevier Science Ltd.
All rights of reproduction in any form reserved.
ISBN: 0-7623-0718-8

While data show that women are under-represented in college and university faculties and that their compensation, on the average, is lower than that of men, this does not in itself establish that they are the objects of discrimination, nor does it indicate the extent of discriminatory practices. In order to demonstrate and measure discrimination, rather than merely show differences, studies and analyses are essential (Scott, E. L., 1977: p. vii).

Scott's (1977) publication provided a recommended method, a statistical method, for demonstrating and measuring gender-based salary discrimination between male and female faculty in an academic institutional setting. "The basic technique is to use a regression analysis of the salaries of white male faculty to predict what the salaries of women and minority faculty would be were their characteristics evaluated in the same way as those of their white male colleagues" (Gray & Scott, 1980: p. 174).

For a given group, the estimation equation for predicting salary is a linear combination of year of birth, year of highest degree, level of highest degree, and (if there are several units that have been grouped) the unit indicator . . . This is equivalent to Multivariate linear regression with salary as the dependent variable and the three (or four) variables – year of birth, year of highest degree, level of highest degree, and possibly unit in which located – as the independent variable (Scott, 1977: p. 3).

The librarian professorate, whose primary functions were not teaching, presented special problems and were not included in the group studied by Scott (1977). Consequently, unlike other faculty groups, the librarian professorate basically only had studies which showed differences between average salaries of male and female faculty members. Librarian salary surveys showed a well-established pattern of women, on average, having lower compensation than men (Hildenbrand, 1997; Kyrillidou & Maxwell, 1995; Looker, 1993). In effect, the librarian professorate lacked what the Scott's (1977) recommended methodology offered other faculty groups, a practical, cost-effective way of demonstrating and measuring gender-based wage discrimination between male and female faculty in an academic institutional setting.

The same argument made by Scott (1977) for other faculty groups could be made for librarian professorate. It can be argued that identifying and verifying the existence of salary differences between male and female librarian professorate were not sufficient to "demonstrate and measure discrimination" (Scott, 1977: p. vii). The problem to be solved for the librarian professorate – those librarians who work in an academic library setting and are subject to similar faculty personnel practices and procedures as other faculty groups at their institution – was to find a practical and cost-effective way to demonstrate and measure gender-based salary discrimination. Two salient characteristics of the librarian professorate distinguish them from other groups of academic employees. First, they contributed to the salary decision-making process by making recommendations on merit raises, tenure and promotion as do other faculty. Second, historically the librarian professorate has been considered to be part of a female-dominated profession, much like a nursing faculty.

While there have been a limited number of statistical studies using multiple regression techniques in studying librarians' salaries (Dowell, 1988; Kim, 1980), they focused primarily on identifying which variables, including gender, affect academic librarians' salary levels. However, unlike the Scott (1977) study, they did not provide a cost-effective, practical methodology for flagging individuals in a group and estimating how much was needed to correct for salary inequities based on gender discrimination. The need for such a methodology is especially critical at this time for librarians in which "there remains a puzzling persistence of inequity in a profession that is more than 80% women" (Hildenbrand, 1997) and where current salary survey data suggests a disparity in the average salaries between male and female librarians that has grown from a 1–2% gap to a 5.3% gap advantaging males (Carson, 1996).

A question explored in this study is whether the Scott (1977) study's "recommended methodology" for teaching faculty was applicable, with or without modifications, to librarians who had faculty rank or status and who, for purposes of personnel actions, were considered and treated as regular faculty at their institutions. For the purpose of discussion, this group was labeled "the librarian professorate". Instead of being evaluated on teaching, scholarly activity, and service, the librarian professorate typically is evaluated on librarianship, scholarly activity, and service. However, they are faculty-like in that they tend to follow the same processes as other faculty groups at their institution for merit and salary determinations. If the Scott model is applicable, then this segment of the professorate, the librarian professorate, would have a simple, practical, and low-cost method for demonstrating and measuring sex-based wage discrimination.

PURPOSE OF THE STUDY

There were three basic objectives for this study. The first objective was to test the recommended methodology as described in the Scott (1977) study to determine whether it was applicable without modification as a model to demonstrate and measure sex-based salary discrimination for the librarian professorate. The second objective was to see if the Scott model's predictive capacity could be strengthened by adding additional predictor variables. The third objective was to see if the Scott model could be strengthened by changing the comparison groups. In addition to looking at the salaries of women and minorities in comparison with white males, as was done in the Scott (1977) study, a comparison of women's salaries with male's salaries was also carried out.

Hypotheses

In this study the following hypotheses were to be tested:

(1) The recommended methodology for demonstrating and measuring sex-based wage discrimination in the Scott study, i.e. predicted regression that is applicable for "regular" faculty, is applicable without any modification for the librarian professorate.
(2) The precision of Scott's model is significantly strengthened by including additional predictors.
(3) The precision of Scott's model is strengthened by comparing the salaries of all males against all females.

First, the individuals included in the study are limited to those in the librarian professorate who have regular, full-time appointments and were employed as of January 1, 1997, at the five largest, public state-supported academic libraries in Illinois. These include Illinois State University (ISU), Northern Illinois University (NIU), Southern Illinois University-Carbondale (SIUC), University of Illinois-Chicago (UIC), and University of Illinois-Urbana/Champaign (UIUC). Each of these academic libraries has a librarian professorate of sufficient size to assure that the group studied is large enough to obtain a reasonable estimation and meets or is close to meeting the Scott (1977) study's recommendation of "at least fifteen white males" (Scott, 1977). In addition, all of the above libraries have librarians who are subject to similar faculty personnel procedures and practices as other faculty groups on their respective campuses. The seven other public state-supported academic libraries in Illinois (Chicago State University, Eastern Illinois University, Governors State University, Northeastern Illinois University, Southern Illinois University-Edwardsville, University of Illinois-Springfield and Western Illinois University) were excluded from this study because the size of their librarian professorate was not large enough to support such a study. (Titus, 1997).

Second, while there are over 184 academic libraries in Illinois (Titus, 1997), access to salary information on academic librarians is not easily obtainable, especially in the private sector. Consequently, this study included only state-supported academic libraries in Illinois where employee salary information is considered public information and was accessible.

LITERATURE REVIEW

There is a substantial body of literature on the topic of gender-based wage discrimination, and for purposes of discussion, three major areas of the literature on this topic were reviewed. First, the literature on the existence of a wage-gap problem was explored from a global perspective, from the perspective of its existence in academe, and from the perspective of its existence in the field of librarianship. The literature on the wage-gap problem was specifically explored to gain a better understanding of the nature and the extent of the wage-gap problem from the above perspectives, especially for the group being studied, the librarian professorate.

Second, the literature on the theoretical frameworks in which scholars attempted to explain the reasons why there were gender-based wage gaps was reviewed. This review provided insights into the scholarly debates and controversies on the major theories associated with discussions on gender-based wage discrimination. A number of theories attempted to explain reasons for gender-based wage gaps. Neo-classical analysis of discrimination, human capital theory, crowding theory, the institutional choice model, the voluntary choice model, and comparable worth theory were all theoretical frameworks which attempted to explain gender-based wage gaps and are included as part of the discussions regarding theoretical frameworks below.

Third, the literature on the use of multivariate regression as a method for determining and measuring gender-based wage disparities was reviewed. This review provided a better understanding of this statistical technique and the extent to which multiple regression techniques were used and accepted in academe as proof of gender-based wage discrimination in the courts in cases involving academic institutions and in the field of librarianship to determine levels of gender-based wage disparities.

The Gender-Based Wage-Gap Problem: Global Overview

The first area to be discussed and explored under the related review of the literature is the wage-gap problem from a global perspective, from the perspective of its existence in academe, and then from the perspective of its existence in the field of librarianship.

Findings generated by empirical studies and research done on wage disparities between men and women have been highly consistent. There has been a longstanding gender-based wage-gap problem (Becker & Lindsay, 1995; Fudge & McDermott, 1991; Milkman & Townsley, 1994; Paul, 1988; Willborn, 1986). "Historically, women have not only earned significantly less money than men, but have also have been employed in what have come to be understood as women's jobs" (Fudge & McDermott, 1991: p. 3). Milkman and Townsley (1994) describe this phenomenon of women primarily being employed in women's jobs as "gender stratification in the labor market" (p. 604).

In addition, national statistics being reported on wages have provided evidence of the longstanding presence of a wage-gap problem. "The wage-gap is a statistical indicator often used as an index of the status of women's earnings relative to men's" (Wage-gap, 1996). Since the mid-fifties, national statistics published by the U.S. Census Bureau on the median earnings of year-round, full-time workers by sex have monitored the extent of the wage-gap problem in the U.S. (DeNavas, 1997; Sorensen, 1991). For over four decades, women's earnings relative to men's have statistically shown there is a serious wage-gap problem. In looking at the most recent wage-gap data being reported, "The U.S. Bureau of Labor Statistics (Dept. of Labor) reports that in 1997, women's weekly earnings were 74% of men's weekly earnings" (National Committee on Pay Equity, 1998).

While some researchers have suggested the wage gap has grown smaller, an analysis of median earnings statistics has indicated "recent increases in the female-to-male earnings ratio have been due more to declines in the earnings of men than to increases in the earnings of women" (DeNavas, 1997: p. ix). Other scholars also have suggested that this is the case (Milkman & Townsley, 1994; Sorensen, 1991).

The Gender-Based Wage-Gap Problem: In Academe

Within academe, the gender-based wage problem also exists, and, as is the case globally, the problem has been longstanding. One of the more recent studies done by Ashraf (1996) indicated that, "in addition to updating previous estimates of the gender differences in faculty salaries, this study allows an observation of the trend in this gap over a relatively significant period, rather than the 'snapshot in time' that previous studies represent" (p. 857). Ashraf investigated the wage differences between male and female faculty in the U.S. from 1969 to 1989. His findings suggested that

> the racial earnings gap narrowed considerably more than did the gender earnings differential over this period. A higher male–female earnings gap at the professor level relative to junior ranks suggests that relatively recent entrants in the academic labor market face a less discriminatory environment than did their predecessors. A controversial finding is that the 'discriminatory component of the earnings differential between male and female faculty fell between 1969–1984, but rose thereafter' (Ashraf, 1996: p. 857).

Earlier empirical studies on the gender wage-gap problem in academia done by Katz (1973), Loeb and Ferber (1971), Johnson and Stafford (1974), and Ferber and Kordick (1978) also have provided a pattern of evidence that suggests women in academe have experienced gender-based wage discrimination. Katz (1973), in studying faculty salaries, concluded that even when female faculty's lower productivity was taken into account, they were still paid less than male faculty. Loeb and Ferber (1971), in taking both numbers of publications and prior experience into account, indicated levels of salary and rank were tied to gender. Johnson and Stafford (1974) concluded, "The most important finding is that the academic salaries of females start out not much less than those of males (4 to 11% less in the six disciplines in our sample) and then decline in a fairly substantial differential after a number of years of potential experience (13 to 23% at 15 years after the completion of the doctorate)" (p. 901). Ferber and Kordick (1978) concluded women Ph.D.s earn less than their male colleagues and that sex discrimination, rather than "their decision to accumulate less human capital" (p. 238), explained the wage gap.

The Gender-Based Wage-Gap Problem: In the Field of Librarianship

There is also substantial evidence in the library and information science literature suggesting that a wage gap not only exists between males and females in the library profession but has persisted for several decades. For example, the Association of Research Libraries (ARL) has conducted and published the *ARL Annual Salary Survey* since 1973. "The most important series of data on internal pay equity in university libraries began to appear in 1976/77 when the ARL expanded its *Annual Survey* to include breakdowns by sex and position ... " (Ray & Rubin, 1987: pp. 39–40). This report is one of the few salary surveys that currently provides gender- based salary data specifically for academic librarians on an annual basis. For 1997–1998, "the overall gender balance in the 110 Canadian and U.S. university libraries (including law and medical) is 35.22% male and 64.78% female" (ARL, 1998: p. 8). This report also provides evidence on gender-based salary differentials. "Looking at the 1997–98 data ... average salaries for men in most cases surpass those of women in the same job category ... Moreover, the overall salary for women is still only 93% that of men" (ARL, 1998: p. 9). This salary study also "reveals that experience differentials between men and women cannot fully account for all the salary differentials . . . further reveals that the average salary for men is consistently higher than the average salary for women in every one of the experience cohorts" (Association of Research Libraries, 1998: p. 10).

Also, since 1951, the results of an annual salary survey on recent library science school graduates has been reported in *Library Journal*. Salaries reported are broken down by gender. In 1995, the study concluded salaries, "if you're a man entering the profession, you're likely to be earning on average 5.3% more than your female counterpart, whatever your title. The disparity between women's and men's starting salary appears to be creeping up, after years of hovering in the 1–2% range" (Carson, 1996: p. 29).

Published (one-time or regular schedule) research on salary surveys focusing on specific groups in the field of librarianship such as library directors, special librarians, M. L. S. graduates, etc., show similar evidence of the gender wage-gap problem (Harris, Monk & Austin, 1986; Jones, 1987; Sayer, 1996). Special one-time studies on librarians' salaries looking at the question of gender-based wage gaps also indicate disparities between males and females. In a recent study done by St. Lifer (1994), it was found that "female librarians are still earning less than their male counterparts by an average of almost 10%. . . . Looking at the disparity in pay between sexes by library sector, women have narrowed the gap in special libraries, where they trail men by just 4.5% ($39,830 vs. $41,660). However, women make a full 13% less ($35,000 vs. $40,700) than men in academic libraries and 7% less than men in public libraries ($34,250 vs. $36,860)" (St. Lifer, 1994: p. 46).

Clearly, the above provides solid and sustained evidence of the existence of a gender-based wage gap for librarians in general and specifically for those in the librarian professorate. More recent studies indicated not only the presence of a sustained gender-based wage gap but also that the gender-based wage gap for the librarian professorate is considerably larger than that for librarians in other types of libraries. Also, the level of disparity between male librarians and females librarians appeared to be increasing instead of narrowing. These findings parallel those suggested by Sorensen (1991) and DeNavas (1997).

In summary, the review of the literature, whether from a global perspective, from the perspective in academe, or from the perspective of the field of librarianship on the extent of the gender-based wage-gap problem, was consistent. There was a wage-gap problem, and historically the wage-gap problem had been longstanding. Within the field of librarianship, not only was there a wage-gap problem for all librarians but there were also differences in the magnitude of the wage-gap problem by type of library. Academic libraries had a greater wage-gap problem than other types of libraries.

Theoretical Frameworks: Overview of Major Theories Explaining
Reasons For Gender-Based Wage Gaps

The second area of discussion and exploration of the related literature was on some of the major theoretical frameworks through which scholars attempted to explain the reasons why there were gender-based wage gaps. The discussion below is broken into two parts. In the first part, an overview of major theories is given. In the second part, a review of the related literature on the concept of pay equity, the principle of equal pay for equal work, and comparable worth theory are discussed.

A review of the literature in several disciplines including economics, higher education, library science, sociology, and women's studies identified several major theories which attempted to explain why there are wage differences between males and females. In the overview, some of these major theories are reviewed. Following this overview, a brief discussion and review of the related literature on pay equity, equal pay for equal work, and comparable worth theory is done. Comparable worth theory is one of the more contemporary theories being discussed and debated on why there are wage disparities between males and females.

While there appeared to be general agreement that gender-based wage-gap problems existed, the theoretical approaches which attempted to explain reasons for gender-based wage-gap problems did not agree. Not only were there different theoretical approaches which attempted to explain reasons for gender-based wage gaps, but there also appeared to be different analytical approaches and "notions about the nature and persistence of discrimination" (England, 1992: p. 117). Blau and Kahn (1994) indicated that, "When analyzing gender differentials in pay, economists commonly focus on male-female differences in skills and on differences in the treatment of equally qualified men and women (i.e. discrimination). Both of these may be considered *gender-specific* factors influencing the pay gap" (p. 23). Another perspective on the way economists approached their analysis of gender differentials was suggested by Willborn (1986). He noted that two types of analysis prevailed. "The first focuses on differences between male and female workers and the second on differences between male and female jobs" (p. 5).

In the late 1950s, Gary Becker did an analysis of male-female differences in skills, i.e. productivity. He was considered "the first to provide a Neo-classical analysis of discrimination in labour markets" (Blaug & Sturges, 1983). He proposed a theoretical framework "for analyzing discrimination in the market place because of race, religion, sex, color, social class, personality, or other non-pecuniary considerations" (Becker, 1971: p. 153). For Becker, discrimination existed when wage decisions were based on factors other than productivity.

> Thus, in theory, one should be able to determine if the wage disparity between men and women is caused by discrimination by comparing the productivity of men and women. If men are sufficiently more productive than women to account for the wage disparity, it is not the result of discrimination. If, however, productivity differences do not account for the entire wage disparity, the disparity not accounted for is attributable to discrimination (Willborn, 1986: p. 10).

Willborn (1986) argued researchers had difficulties when they attempted to implement Becker's theory. He indicated that the direct measurement of wages and productivity presented difficult implementation problems for researchers which resulted in other theoretical approaches being tried.

Instead of attempting to measure productivity, directly then, researchers have attempted to measure it indirectly by using worker characteristics such as education, health, and experience as proxies for productivity. As a result, in practice the research has compared male and female wages to characteristics of male and female workers. To the extent worker characteristics do not account for wage differentials, the residual wage differential is attributed to discrimination (Willborn, 1986: p. 11).

This research approach was based on human capital theory. Human capital is defined as, "the stock of acquired talents, skills, and knowledge, which may enhance a worker's earning power in the labor market" (Kuper & Kuper, 1996: p. 376). "Under human capital theory workers invest in their earnings power or human capital by acquiring those characteristics that are valued in the labor market" (Willborn, 1986: p. 11). In effect, the level of wages are tied to the amount of human capital investments made by an individual. Human capital theory argued that the reasons for differences in men's and women's salaries was not because of sex discrimination, but were based instead on differences in the levels of investment in human capital.

For researchers using the human capital theory approach, such as Fuchs (1988), Polachek (1981), and Zellner (1975) , findings indicated "that women's own occupational choices and lesser human capital investments generally understood as motivated by commitments to child rearing (themselves exogenous to the model) are the main mechanisms producing occupational segregation and wage differentials by gender" (Milkman & Townsley, 1994: p. 611).

However, as was the case with Becker, there were serious limitations with the human capital theory approach. Willborn (1986) summarized a number of the criticisms pertaining to human capital theory studies, which included a lack of universal agreement on the definition of discrimination, that productivity was just one of several factors influencing wages, that there were limits to multiple regression techniques used regarding worker characteristics and earnings, that defining worker characteristics was difficult, and that human capital studies resulted in underestimating the impact of sex discrimination on wages (p. 16).

Another major criticism of human capital theory was that the human capital empirical studies that were done left too much of the differences in the wage gap between men and women unexplained. It could be argued that what was unexplained was attributable to gender-based wage discrimination. "In the human capital studies that have been done including Blinder (1973), Corcoran (1979), Mellor (1984), and Rytina (1982), worker characteristics account for only a small portion of the earnings difference between men and women" (Willborn, 1986: p. 12). Willborn (1986) indicated the best of the human capital studies explained less than 50% of the wage gap between men and women. Milkman and Townsley (1994) also were critical of human capital theory. Based on the research findings of England (1982); England, Chassie and McCormack (1982); and England, Farkas, Kilbourne and Dou (1988), they argued that "human capital theory is largely inconsistent with the available empirical evidence" (Milkman & Townsley, 1994: p. 611).

Instead of studying workers' skills and productivity or workers' characteristics, some scholars took a different approach and investigated occupational segregation and the extent it contributed to male-female wage gaps (Bergmann, 1986; England, 1992; Sorensen, 1989; Willborn, 1986). These scholars focused on an analysis of the differences between male and female jobs. "Two discrimination theories postulate why occupational segregation contributes to the male-female earnings gap; the crowding hypothesis and the institutional approach" (Sorensen, 1989). The crowding theory suggested discrimination in the labor market was the result of occupational segregation (Stevenson, 1975). There was men's work and there was women's work.

> Employers discriminate against women by excluding them from occupations considered to be 'men's work'. Since these jobs are reserved for men, relatively few women are hired into these positions. Given that the demand for women in these jobs is limited, they are crowded into other occupations, typically referred to as 'women's work'. The supply of women accordingly increases for 'women's work', which in turn reduces their wages. For simplification, this model assumes that women and men have equal abilities, and without discrimination would be equally paid (Sorensen, 1994: p. 48).

The institutional model of discrimination stated that some types of firms had internal labor markets. Instead of wages being set by supply and demand, wages were set by local rules and customs. "According to the institutional model of discrimination, firms with internal labor markets are more likely to discriminate than other firms" (Sorensen, 1994: p. 47). This was because they reflected societal norms in which women's work was less valued. "Both [models] argue that employer discrimination contributes to occupational segregation and the wage penalty against 'women's work' (Sorensen, 1994: p. 56).

Sorensen (1994) also criticized the proponents of voluntary choice models, the human capital model and the theory of compensating differentials, arguing that these voluntary choice models do not explain the presence of a severe wage penalty for women. "Only discrimination theories predict a wage differential between female and male-dominated jobs once productivity characteristics have been taken into account" (p. 55). The counterargument by proponents of voluntary choice models was "that these studies used a mispecified model. According to this view, insufficient variables have been included in these analyses to control for productivity and supply differences between men and women"

(Sorensen, 1994: p. 55). Sorensen's rebuttal was that even when researchers made serious efforts to correct for a mispecified model, "the estimated coefficient for the sex composition of an occupation remains significant. Thus, the voluntary choice models and their adherents have not adequately explained this severe wage penalty" (Sorensen, 1994: p. 55). Sorensen's conclusions from her research findings on the nature of discrimination women experienced were, "I conclude that this underpayment of 'women's work' results from economic discrimination of women. . . . Economic discrimination exists when workers of one sex are denied economic opportunities available to workers of another sex for reasons that have little to do with their individual abilities" (Sorensen, 1994: p. 130).

Pay Equity, Equal Pay for Equal Work, and Comparable Worth Theory

Another theory which attempted to explain gender-based wage gaps was comparable worth theory. This theory is based on the concept of comparable worth, defined as "equal productivity; the basis for paying workers the same" (Rutherford, 1992: p. 84). It was viewed as a way to counter gender-based wage discrimination (Rutherford, 1992: p. 84). Comparable worth was also seen as "a shorthand term for equal pay for jobs of comparable worth: the proposition that 'comparable jobs' should receive the same pay" (Hill & Killingworth, 1989: p. 1). Comparable worth theory was not considered by many scholars a new idea, but an idea with its early origins in the 1900s that did not really take hold until the 1980s (Blum, 1991; Feldberg, 1980; Paul, 1989). Feldberg (1980) indicated women in the Bookkeepers and Accountants Union No. 1 of New York provided an example of comparable worth's early origins when they compared their wages to higher paid men who were hod carriers (p. 57). Paul (1989) stated, "In fact, the notion of 'comparable work' was employed by the National War Labor Board during World War II. The board required equal pay for 'comparable work' and made job evaluations within plants between dissimilar jobs to determine whether any pay inequities existed" (p. 1). Also, comparable worth theory appeared to have emerged and had been an outgrowth from other theories, ideas, and concepts. Specifically, comparable worth theory seemed influenced by the concepts of pay equity and equal pay for equal work.

In the literature, pay equity as it relates to comparable worth was discussed in two different contexts. In the first context, pay equity and comparable worth were considered distinctive in meaning from one another, but interrelated. For many scholars (Fudge & McDermott, 1991; Paul, 1988, Willborn, 1986), pay equity did not have the same meaning as comparable worth. Fudge and McDermott (1991) argued pay equity was goal directed. The goal was to end wage discrimination (p. 4). "Pay equity was also based on the assumption that women do valuable work – work that is, in most cases, as valuable to society and employers as work done by their male co-workers" (Fudge & McDermott, 1991: p. 4–5). In contrast, comparable worth provided a process for ending wage discrimination. Willborn (1986) also viewed pay equity and comparable worth as two distinct entities. He argued, "The term comparable worth, properly used, refers only to the effect of an occupation's sex composition on the compensation of that occupation. Pay equity refers to any wage disparities that may be considered inequitable" (Willborn, 1986: p. 3). In general, pay equity was viewed as a more generalized, broadly applied term. Comparable worth was viewed as a narrower, more specific term, a subset of pay equity.

For others, the terms "pay equity" and "comparable worth" were sometimes used interchangeably (Hill & Killingsworth, 1989). For example, Paul (1988) suggested, "Comparable worth, or pay equity in its newest guise, is an attractive concept: if only employers could be required to pay female employees in traditionally female occupations the same salaries as males in male-dominated jobs of comparable value to their employers, then the 'wage-gap' would largely disappear" (p. 1). However, of the two different contexts used, treating pay equity and comparable worth as distinctive and different in meaning appeared to be the most commonly used approach in the literature.

It also appeared that comparable worth theory was influenced by the ideas associated with equal pay for equal work. Blum (1991) argued that the concept of equal pay for equal work influenced the thinking of proponents of comparable worth theory. One built on the other. "The rationale for comparable worth stems from the widely accepted tenant of 'equal pay for equal work': that men and women with the same jobs should receive the same pay" (Blum, 1991: p. 4).

As suggested previously, it was not until post-World War II and after that comparable worth theory gained momentum. It was argued that, in part, this "gained momentum" for comparable worth theories was based on the failure of the Equal Pay Act of 1963 to solve the wage-gap problem. It was assumed that if men and women received the same pay for the same work, wage-disparities based on sex discrimination would be resolved. However, "ironically, the equal pay acts, rather than correcting the male-female wage disparity, merely highlighted the subtlety of the problem" (Willborn, 1986: p. 1). Until the 1960s, "it was common practice for employers to pay women less than men even for the same job" (Willborn, 1986: p. 1). Then the Equal Pay Act of 1963 was passed. The act made it illegal for employers to pay women less than men if they were doing the same job. However, according to its critics, this did not solve the wage-disparity problem. Willborn (1986) indicated, "The equal pay acts are now almost universally observed and women generally receive the same pay as men doing the same work. Nevertheless, women continue to earn, on average, about 60% of what men earn. Why? Because unequal pay for equal work was not the principal cause of the male-female wage disparity. Other more complex factors were at work" (p. 1). Paul (1988) also argued that the Equal Pay Act of 1963 did not go far enough. "It leaves important gaps in the protection afforded to women workers; for example, women who labor in jobs with no equivalent male jobs available for comparison are left unprotected as are women whose work is comparable to men's but not equal by the 'substantially equal' standard" (p. 81). Blum (1991) pragmatically pointed out that, "Because few women have the same jobs as men, however, this popular principle does little to advance women's interests" (p. 4).

As stated earlier, comparable worth theory, much like other discrimination theories discussed earlier, attempted to provide yet another explanation for the wage-gap problem. The theory, in effect, attempted to get at what was considered by proponents of comparable worth to be the principal cause of the wage-gap problem, occupational segregation. As Sorensen (1994) stated, "Across the country, women working as nurses, librarians, and secretaries argue that their jobs are paid less than jobs of comparable value held primarily by men that is jobs requiring comparable skill, effort, responsibility, and working conditions" (p. 3). The remedy to correct wage disparities between "women's work" and "men's work" was found in comparable worth theory.

Many scholars (Blum, 1991; Hill & Killingsworth, 1989; Sorensen, 1994; Willborn, 1986) appeared to be in general agreement on the assumptions associated with comparable worth. They held that

> Proponents of comparable worth argue that extensive occupational segregation is a major cause of the earnings disparity between women and men. They argue that because of occupational segregation, certain jobs become identified as "women's work". This label results in lower pay, simply because women do the work. Thus, "women's work" is underpaid and this underpayment is a principal reason why women earn less than men (Sorensen, 1994: p.7).

Comparable worth was seen as a way to correct for wage disparities between males and females which resulted from occupational segregation and the devaluation of "women's work". Simply put, "the doctrine [comparable worth] simply states that an employer should pay employees in jobs held predominately by women the same as employees in jobs held predominately by men if they require comparable skills, effort, responsibility, and working conditions" (Sorensen, 1994: p. 3).

Arguments of opponents or skeptics of comparable worth fell into a few thematic categories. Economic arguments made against comparable worth represented one of these categories. Paul (1989), in making a case against comparable worth, summarized many of the economic arguments represented in the literature. "The main charges against comparable worth can be stated succinctly: that the market is not inherently discriminatory; that the asserted wage-gap of 40% is grossly inflated; that it would be too expensive to implement; that it would disrupt the U.S. economy by increasing inflation, driving up unemployment, and making U.S. products less competitive on world markets; that it would harm women's employment by overpricing their services; that it would hurt blue-collar workers because comparable worth evaluation schemes favor education over manual labor; and finally, that reading such a standard into current antidiscrimination legislation would penalize employers for wage-setting practices over which they have little control" (Paul, 1989: p. 39). Also, the proponents of a market-based economic system did not want their system fundamentally changed; "the opponents see only a radical, if not revolutionary, half-baked scheme to overturn our market economy" (Paul, 1989: p. 57). Other proponents of a market-based economic system argued the market would, in time, eventually self-correct. "Thus, comparable worth, which they see as a radical departure from the United States' free-market heritage, would soon be as unnecessary as it is now undesirable" (Paul, 1989: p. 39). Simply put, neoclassical economic theory conceded economic discrimination existed, but economic discrimination was viewed as a temporary phenomenon (Hutner, 1986).

Arguments made by critics on the viability of using comparable worth theory to award damages by courts represented another thematic category. Sorensen (1994) and Willborn (1986) argued that there was no agreement in the courts as to their acceptance of comparable worth as a legal theory. Willborn (1986) observed, "Other courts have split: some seem to accept comparable worth as a viable legal theory, while others reject it with unequivocal language" (p. 33). It was argued that, even though comparable worth advocates made some inroads into comparable worth challenges with the Supreme Court's decision in *Gunther v. Washington*, these gains have not been great. "For despite the seeming room left open by the Supreme Court's decision in *Gunther v. Washington*, other courts have not been very receptive to cases brought under a comparable worth interpretation of Title VII. In fact, *Gunther*, has not really changed the courts' attitudes much on this score" (Paul, 1989: p. 99). Others argued that the legal process was too slow to be considered a viable, cost-effective, solution for correcting the gender-based wage-gap problem (Raymond, Sesnowitz, & Williams, 1990). Court cases took years to work their way through the system. For example, in Canada, "in a landmark decision, a human rights tribunal ruled Wednesday [July 29, 1998] that public servants in female-dominated job categories deserve compensation for years of unequal pay" ("Females workers win salary dispute", 1998: p. 14). This decision was the culmination of legal action which was initiated fourteen years earlier in 1984.

In summary, based on the above, discussions and debate on several theories which attempted to explain wage-gap problems were well represented in the research literature. More recent discussions and debates centered on the pros and cons of comparable worth theory. The theories reviewed and described above have all attempted to explain why there were gender-based wage gaps. All of the theories had limitations, and some were better at explaining part of the reasons that there were gender-based wage gaps. However, none of them totally explained the disparities in wages between males and females. What could not be explained was considered to be attributable to sex discrimination. Willborn (1986) commented that "the human capital studies that have included occupational characteristics, as an explanatory variable, have generally explained more of the male-female wage disparity than studies that have included only worker characteristics" (p. 25). In addition, while he suggested wage disparity was better explained by studies based on crowding theories, these studies were also not adequate. "In summary, the studies that have been focused on occupational segregation as the explanation for the male-female wage disparity also fail to resolve the discrimination issue. Although the disparity is better explained by occupation and worker characteristics alone, the studies cannot account for the total disparity and any residual may be attributable to discrimination. Even if the studies could account for the total disparity, sex discrimination could not be ruled out as a causative factor because it may be that occupational segregation itself is caused by discrimination" (Willborn, 1986: p. 29).

Methods for Determining Wage Disparities: Multivariate Regression in Academe

The third area of discussion and exploration of the related literature related to the extent to which multiple regression could be used as a method for determining and measuring gender-based wage disparities in academe and in the field of librarianship. Also discussed was the level of acceptance of multiple regression statistical techniques as valid and acceptable proof of evidence of gender-based wage discrimination cases involving academic institutions in the courts.

Within academe, the use of multiple regression as a way to identify gender-based wage discrimination began to take hold in and to be reported on in the literature in the seventies. Regression analysis was used to determine if pay inequities based on gender existed and to estimate the magnitude of salary disparities. "Equity pay strategies generally use regression analysis to determine a base target regression line, which is usually the male-dominated one. Adjustments are made in relation to that line" (Moore & Abraham, 1995).

Bergmann and Maxfield (1975) in their study of faculty salaries at the University of Maryland-College Park provided "a prototype of a study which could be done fairly easily by a faculty group on any campus where the salary information was available" (p. 262). By the late 1970s, a number of academic institutions had published or circulated results of their research studies on the salaries of men and women on their campuses. Many of these campus salary studies used multiple regression techniques as their research method (Raymond, Sesnowitz & Williams, 1990: p. 197). The use of multiple regression techniques in academe to measure gender-based wage discrimination further gained momentum with Scott's publication in 1977, *Higher Education Salary Evaluation Kit: a Recommended Method for Flagging Women and Minority Persons for Whom There is Apparent Salary Inequity and a Comparison of Results and Costs of Several Suggested Methods*, published by the American Association of University Professors.

By the 1980s and 1990s, the use of multiple regression to measure salary disparities between male and female faculty were routinely being reported. While in the past there had been some debate surrounding what methods. (e.g, paired-comparison method, actuarial method using multiple regression, etc.) should be given credence in recognizing systemic salary discrimination, Dean and Clifton (1994) argued, "At present, there appears to be some consensus developing in the published literature about how to conduct pay equity research" (p. 113). For example, in conducting pay equity research, multiple regression and other statistical techniques are frequently used and considered by many pay equity researchers as an acceptable, frequently used, research methodology (Bergmann, 1975; Dean & Clifton, 1994; Paul, 1989; Raymond, Sesnowitz & Williams, 1990). Moore (1993) in her study which did an evaluation of salary equity models, concluded, "A regression model with a sex variable is the most straightforward and most easily understood method of analyzing for salary equity" (p. iii). She further argued that the choice of a regression method was not as important as choosing what variables needed to be included in the salary equity model. "The most critical decision element in designing a salary equity model is deciding which variables to include" (Moore, 1993: p. iii). Other researchers also concluded that the choice of variables to be used in the regression analysis were important. Dean and Clifton (1994) in their evaluation of pay equity studies done at five Canadian universities indicated that, while the number of variables included and the way the variables were measured varied for the universities studied, all used multiple regression analysis (p. 89).

Coupled with the trends in the literature, which suggested growing acceptance of the use of multiple regression, were ongoing discussions and debates on the strengths, the weaknesses, and the limitations associated with using these techniques (Pezzulo & Brittingham, 1979; Raymond, Sesnowitz & Williams, 1990). Some of the strengths associated with multiple regression and discussed in the literature included its flexibility. "According to Kieft (1974), its attractiveness lies in its flexibility; it can be adapted to any employee group at an institution without substantial modification of the methodology" (Pezzullo & Brittingham, 1979: p. 7). Most frequently discussed in the literature on weaknesses or cautions associated with using multiple regression analysis included the need to be cautious in interpreting results (Pezzullo & Brittingham, 1979), the questioning of the ability of multiple regression analysis to identify both ethnic and sex discrimination (Raymond, Sesnowitz & Williams, 1990; Sowell, 1984), and the need to address statistical design problems associated with multiple regression such as unreliability, multicollinearity, and heteroscedasticity (Willborn, 1986: pp. 12–16).

Also, discussed in the literature on regression analysis was the question of whether the courts, in making decisions on discrimination cases, would accept statistical evidence such as regression analysis as proof of gender-based wage discrimination. Sobel (1977), Spaulding (1984), Bazemore (1986), and Raymond, Sesnowitz and Williams (1990) all discussed how various courts in their decisions on cases had validated in their court decisions the use of regression analysis as a form of proof of discrimination. In the case of *Sobel v. Yeshiva University* (1977), the court as part of its opinion included a general discussion on the use of statistical proof in employment discrimination cases. "Such a heavy reliance upon statistical evidence in employment discrimination cases has, of course, been widely accepted and even received the imprimatur of the Supreme Court" (p. 161). In 1984, in the case of *Spaulding v. University of Washington* (1984) , "the Ninth Circuit Court of Appeals was willing not only to reject the 'matched pairs' or 'comparator' approach, but went on to imply strongly, that the plaintiffs should have used regression analysis" (Raymond, Sesnowitz & Williams, 1990: p. 197). Then in 1986, the United States Supreme Court in *Bazemore v. Friday* not only acknowledged the legitimate use of regression analysis to provide statistical evidence in discrimination cases but also stated, "Regression analysis that includes less than 'all measurable variables' may serve to prove a Title VII claim, since preponderance of evidence, not scientific certainty, is standard of proof" (*Bazemore v. Friday*, 1986, 6.Statistics-108.8125).

Methods for Determining Wage Disparities: Multivariate Regression in the Field of Librarianship

While the use of regression techniques as a means to demonstrate and measure salary discrimination among teaching faculty appeared to be very prevalent in higher education, there was less evidence of its application and use as a tool for analysis to determine gender-based salary discrimination in librarians' salaries. In part, this lack of evidence in the library literature was because the library professorate were frequently excluded from faculty salary studies done on their campuses. (Scott, 1977; Scott et al., 1993). In part, it appeared that it just took longer for the use of regression techniques to "take hold" in the field of librarianship. In 1987, the American Library Association (ALA) in its T. I. P. (Topics in Personnel) KIT series provided librarians with general information on the use of multiple regression analysis and suggested, "It is a technique which has worked well for other professionals, and it could prove very useful for librarians" (*Pay Equity*, 1987, Chap. Strategies).

In the library-related literature, studies using regression techniques to examine the question of which variables influence academic librarians' salaries are not as numerous as the studies being reported on faculty in academe. However, some studies were reported in the late 1970s and the 1980s (Dowell, 1986; Genova et al., 1977; Kim, 1980). In these studies, all the investigators' findings suggested factors other than sex discrimination contributed to explaining salary differences between male and female librarians' salaries. However, when these factors were held constant, salary differences remained that could be attributed to sex differences. While the studies were similar in that they all concluded that factors other than sex contributed to explaining the differences in the salaries of male and female librarians, the investigators drew different conclusions on which factors other than sex explained these differences.

Genova, Gill and Cole (1977) published their findings on an "investigation of salary differentials among State University of New York librarians representing university centers, state colleges, agricultural and technical colleges, and special and medical schools" (p. 1). Their use of multivariate regression analysis indicated wage disparities between salaries of men and women. "Rank was found to be the strongest predictor of salary, other significant predictors being years of professional experience, and sex" (Genova, Gill & Cole, 1977: p. 1).

U. C. Kim (1980) also used multivariate statistical techniques to investigate which variables explain salary differences among academic librarians from medium-sized, state-supported academic institutions located in the Midwest. The interrelationships of 30 variables were studied and "nearly 20 salary predictor variables accounted for approximately 80% of the librarians' salary variance, leaving 20% to be explained" (Kim, 1980: p. 155). Kim's findings identified age, the highest library degree, years of professional library experience, the number of professionals supervised, certain professional activities such as assignments in profession organizations, mobility, and faculty rank held as statistically significant predictors for academic librarian salaries. "The experience variables group is a stronger predictor than any other group in this study, and the group by itself accounts for 43.5% of the librarians' salary variation. Years of professional library experience is a far stronger predictor of salary variance than years of non-library experience" (Kim, 1980: p. 147). Gender as a factor was statistically significant at the 0.05 level and explained 8.55% of the variance in male and female librarians' salaries (Kim, 1980: p. 152). Kim's findings also suggested that, "the amount of salary difference attributable to sex mentioned above increases as the supervisory level or the rank of the librarians gets higher" (Kim, 1980: p. 153).

D. R. Dowell (1986) used multiple regression and path analysis to investigate the question of the relation of salary to sex among librarians at research libraries located in the South Atlantic Census Region. All librarians, including library administrators at the libraries participating in the study, were mailed a survey and asked to respond. Similar to Kim's (1980) study, Dowell, "attempted to determine whether the apparent influence of sex discrimination on librarians' salaries would be shown to be reduced if other factors were considered" (p. 93). Dowell argued, "Only by examining the influence of many variables can we move closer to understanding the real causes for differences between the salaries of women and men librarians. In addition, this sort of examination might help to clarify what needs to be done to achieve full equity" (Dowell, 1988: p. 93). Dowell's findings indicated 20 variables of the 45 he examined were statistically significant and contributed to explaining pay differences between male and female librarians. Of the 20 statistically significant variables, "level of position was the most powerful predictor of salary – alone accounting for almost 60% of all variance" (Dowell, 1988: p. 92). "However, when 20 such factors were held constant, men still earned almost $1,200 more per year than women" (p. 92), or " . . . sex explained 6.3% of the variance in salaries when it was the only information used to differentiate among the salaries paid to individuals in this sample" (p. 96).

In summary, it appears that, in academe, the use of multivariate regression techniques as a method to determine and measure the extent of gender-based wage discrimination between male faculty salaries and female faculty salaries has taken hold and become generally accepted and utilized by many academic institutions who were attempting to correct gender-based wage disparities. Also, over a longer period of time, the courts validated the use of regression analysis as a form of statistical proof in their decisions on cases brought before them dealing with gender-based wage discrimination. However, in the field of librarianship, the use of multivariate regression techniques as a method to determine and measure the extent of gender-based wage discrimination does not appear to have taken hold as was the case in academe. Primarily, multiple regression techniques has been used in studies which focused on studying the influence different variables have on librarians' salaries. Unlike the Scott (1977) study, these studies do not provide a recommended method to determine and measure the extent of gender-based wage discrimination for academic librarians.

METHODOLOGY

Five libraries were included in this study, including the five largest public, state-supported academic libraries in Illinois, Illinois State University (ISU), Northern Illinois University (NIU), Southern Illinois University-Carbondale (SIUC), University of Illinois-Chicago (UIC), and University of Illinois-Urbana/Champaign (UIUC). These five libraries were asked to provide data for all tenure-track, regular, full-time appointments for librarians working in their libraries who were employed as of January 1, 1997. Library faculty who had temporary, part-time, or nontenure-track appointments were not included.

For purposes of analysis, the group of librarians to be studied was defined as "the initiating unit, the group that makes decisions to send an appointment or a promotion to the upper authorities for approval, . . . the group that should be considered in a study of possible salary inequities. In many colleges and universities this unit is the department" (Scott, 1977: p. 2). In most cases, for academic libraries with a librarian professorate, the initiating unit is also the department, with the entire library organization being considered the equivalent of a single academic department. Since the same equation is used to estimate salaries for all members of the group, grouping on the basis of salary-making decisions results in strengthening the homogeneity of the group (Scott, 1977: p. 2). For all the libraries in this study, those included met the above conditions for inclusion or exclusion in the group being studied. All individuals had to be considered part of the same initiating unit. They all were subject to the same salary decision-making processes. There were no missing cases. At each institution, if the conditions for inclusion in the study had been met, all individuals who met the conditions were included in the study.

For librarians who held administrative appointments in the library, their inclusion or exclusion in the study varied. If they were part of the initiating unit, they were included in the study. For example, at NIU, all librarians with administrative appointments, except the Director, fall subject to the same salary decision-making processes as other NIU library professorate and were included in the study. If they were not part of the initiating unit, they were not included in the study. ISU and UIUC, librarians with administrative appointments are not subject to the same salary decision-making processes other library professorate are in their respective libraries and were excluded.

In addition, the size of the group had to be large enough to assure reasonable estimates. Scott (1977) recommended there be at least fifteen white males in the group being studied.

Two types of comparisons within a group were made. The first type compared white males against females and minorities. This was the comparison used in the Scott (1977) study. NIU, UIC, and UIUC had group populations which included white males, females, and minorities. ISU and SIUC did not have group populations which included minorities. Consequently, ISU and SIUC could not be used for this type of group comparison. The second type compared males against females. All five libraries had group populations which included males and females.

Using the two types of comparisons within a group outlined above, five models were tested using procedures as described in the Scott (1977) study's recommended methodology. The five models tested included:

Model 1 (Scott3). This is the basic model that was recommended in the Scott (1977) study. Salary is the dependent variable. The three independent variables are year of birth, year of highest degree, highest degree level, and, if several units are grouped, unit indicator. For this study, none of the five libraries used grouped units. Therefore, Model 1 (Scott3), for purposes of this study, included only the first three independent variables listed above.

Salary=function (year of birth, year of highest degree, highest degree level).

Model 2 (Scott3 plus 3). This model builds upon Model 1 (Scott3) and adds additional predictors. The dependent variable is salary. The six independent variables are year of birth (Scott), year of highest degree (Scott), highest degree level (Scott), level of position (Dowell), rank held (Kim), and years of professional library experience (Kim).

Salary=function (year of birth, year of highest degree, highest degree level, level of position, rank held, and years of professional library experience).

Model 3 (Scott2). This model simplifies Model 1 (Scott3) even further. Salary is the dependent variable. The two independent variables are year of birth and year of highest degree. The model suggests that for the library professorate the highest degree level is predominately the same, a master's degree. That is, there is no variance. Consequently, this variable does not contribute to explaining salary.

Salary =function (year of birth, year of highest degree).

Model 4 (Scott2 plus added predictors). This model builds upon Model 3 (Scott2). The dependent variable is salary. The five independent variables are year of birth (Scott), year of highest degree (Scott), level of position (Dowell), rank held (Kim), and years of professional library experience (Kim).

Salary=function (year of birth, year of highest degree, level of position, rank held, years of professional library experience).

Model 5 (Stepwise/predictors). This model takes statistically significant predictors identified by doing multiple linear STEPWISE regression statistics on the above models. The dependent variable is salary. The three independent variables are year of birth, rank held, and level of position.

Salary=function (year of birth, rank held, and level of position).

The recommended methodology for identifying salary inequities for women and minorities as presented in the Scott (1977) study was followed. For all five libraries participating in the study, a statistical method, multiple linear regression analysis, was used to test all five models comparing white males with females and minorities, if applicable, and males with females of the same group. This statistical method provided an estimate of the dollar amounts needed to bring the actual salaries of females and minorities or females up to comparable white males' salaries.

First, an estimation equation for predicting salaries was calculated.

> For a given group, the estimation equation for predicting salary is a linear combination of year of birth, year of highest degree, level of highest degree, and (if there are several units that have been grouped) the unit indicator. The best fitting linear combination is found by least squares. This is equivalent to multi variate linear regression with salary as the dependent variable and the three (or four) variables – year of birth, year of highest degree, level of highest degree, and possibly unit in which located – as the independent variables (Scott, 1977: p. 3).

Using SPSS 6.1 for Windows, a statistical analysis and database-management software package, a multivariate linear regression procedure was used to calculate a predicted salary estimation equation for white males and males in the two types of comparison groups (males/females and white males/females and minorities). Salary (adjusted) was the dependent variable and the independent variables depended upon which of the five models tested was being used.

For purposes of doing analysis on the predictors being used in the five models, the multivariate linear regression procedures were run using two methods. The default method ENTER was used, as well as the method STEPWISE. In the ENTER method, "Variables in the block are entered in a single step" (Norusis, 1993: p. 358). In the STEPWISE method, "variables in the block are examined at each step for entry or removal" (Norusis, 1993: p. 359).

The results of the multivariate linear regression procedure using the ENTER method which was done for either white males or males were used to calculate a predicted salary estimation for females and minorities or females in the same initiating unit.

"Now the predicted salary of a woman member of the group, if she were paid on the same basis as a white male in the group, would be obtained by substituting the attributes of this woman into the right side of the equation" (Scott, 1977: p. 3).

Then, residuals were computed for females and minorities or females by subtracting the predicted salaries that had been computed from their actual salaries (adjusted),

which is the difference between the actual salary received and what is estimated from the prediction equation for a white male of the same attributes. When this residual is negative, the woman (or minority person) appears to be *underpaid* by the indicated amount and has been flagged" (Scott, 1977: p. 3).

Residuals for females and minorities or females = actual salary (adjusted) − predicted salary estimation.

For this study, the dependent variable was salary. For testing the first hypothesis, the independent variables used in the calculations were actual contracted salaries for those members of the librarian professorate employed as of January 1, 1997, recorded in dollars adjusted to the same base (12 months), year of birth, year of highest degree, and highest degree level. The variable indicator of units (used only if units must be combined) was not used because none of the libraries participating in the study combined units. For testing other hypotheses, in addition to the above independent variables, other independent variables, to include level of position (Dowell, 1986), years of professional library experience (Kim, 1980), and rank held (Kim, 1980) were tested.

Because of local campus policies, UIUC was unable to release information on the variable year of birth. Consequently, for UIUC the variable year of birth was estimated based on a multiple linear regression method using UIC data to obtain year of birth estimated (YRBIRTHE)

Estimation equation for year of birth (YRBIRTHE):

Year of birth=function (year of highest degree, years of previous library experience).

For the five libraries in the study, information was provided by a library administration contact person from each of the libraries. These individuals had access to library faculty personnel data for their respective institutions. Information was obtained from their libraries' personnel and administrative records. Written general instructions on the data to be gathered and data reporting forms were provided and reviewed by the investigator with each library administrative contact participating in the study to insure the accuracy, reliability, and integrity of the data reported. The data gathered on each individual included in the study for each library in the study included the individual's name (optional), institution, actual contracted monthly salary in dollars (rounded), salary in dollars (rounded) adjusted to a 12-month base, year of birth, indicator of highest degree, year of highest degree, gender, whether the person was Caucasian or a minority, level of position, years of professional library experience, and rank held.

To prevent any possible selection bias, data was gathered for all members of the group being studied. Scott (1977) points out, "It is important that all information be collected for each member of the group since any selection is likely to be biased and might bias the estimates"(Scott, 1977: p. 2). ISU, NIU, SIUC, and UIC not only reported data for all members of their group but reported data for all categories requested. UIUC reported data for all members for their group and all categories requested except for year of birth.

NEGATIVE (–) RESIDUAL ANALYSIS

The five models previously described were tested using procedures in the Scott (1977) study's recommended methodology, using a multivariate linear regression procedure. The findings of the tests done on the five models are discussed below. Specifically, the discussion on findings first looks at the number of cases showing negative (–) residuals to determine if salary disparities exist and to what extent they exist. This is a descriptive analysis which provides evidence of salary disparities.

Residuals were calculated for females (or females and minorities) in each data set by taking the difference between the actual salary that was reported and subtracting from this reported salary the salary that was estimated from the prediction equation for males (or white males) with the same attributes. A negative (–) residual indicated that a female (female or minority) appeared to be underpaid by the dollar amount calculated as a residual compared to males (or white males) with the same attributes.

Two comparison groups (males against females and females and minorities against white males) were studied. The findings for each model, by the comparison groups for each model, are discussed below.

Negative (–) Residuals Analysis

Table 1 shows for the five libraries studied (ISU, NIU, SIUC, UIC, UICU) the percentage of females with negative (–) residuals for all models (1–5) in which the comparison group was males against females. On average, all five models had negative (–) residuals or underpaid salaries for over 50% of the

Table 1. Percentage of cases showing negative (−) residuals, Females.

	Model 1 3 predictor variables (Scott3)	Model 2 6 predictor variables (Scott3 plus added predictors)	Model 3 2 predictor variables (Scott2)	Model 4 5 predictor variables (Scott2 plus added predictors)	Model 5 3 predictor variables (Stepwise/ predictors)
ISU	82%	76%	76%	53%	65%
NIU	71%	29%	62%	38%	36%
SIUC	60%	67%	93%	67%	67%
UIC	62%	46%	65%	46%	62%
UIUC	61%	57%	64%	61%	54%
MEANS	67%	55%	72%	53%	57%

females for the five libraries studied. This suggests, regardless of the model used, salary disparities exist at all five libraries studied when males are compared to females.

In addition, it appeared for each library studied that the model used made a difference in the percentage of negative (−) residuals for females for that library. In some cases, the differences in the percentage of negative (−) residuals for a library was marked, while in others, the differences were only slight.

Table 2 shows, for the three libraries (NIU, UIC, UIUC) studied, the percentage of females and minorities with negative (−) residuals for all models (1-5) in which the comparison group was females and minorities against white males. The models in which the comparison group was females and minorities against white males appeared to be more limited in its use than models in which the comparison group was males against females in that it could only be used when data on minorities was reported. Only three (NIU, UIC, UIUC) of the five libraries reported minorities in their data sets. Also, in grouping females and minorities together, gender dispersion became muddied. This was because male minorities become part of the females and minorities comparison group.

On average, all five models had negative (−) residuals or underpaid salaries for 47% or more for the libraries studied. This suggests, regardless of the model used, that salary disparities appear to exist at the three libraries studied where the comparison was between females and minorities and white males. Model 1 had a mean of 61% and the percentage of the negative (−) residuals for the three libraries ranged from 60% to 82%. Model 2 had a mean of 47% and the

Table 2. Percentage of cases showing negative (−) residuals, Females and Minorities.

	Model 1 3 predictor variables (Scott3)	Model 2 6 predictor variables (Scott3 plus added predictors)	Model 3 2 predictor variables (Scott2)	Model 4 5 predictor variables (Scott2 plus added predictors)	Model 5 3 predictor variables (Stepwise/ predictors)
NIU	63%	38%	57%	36%	44%
UIC	63%	48%	63%	48%	65%
UIUC	58%	56%	58%	56%	51%
MEANS	61%	47%	59%	47%	53%

percentage of the negative (−) residuals for the three libraries ranged from 38% to 56%. Model 3 had a mean of 59% and the percentage of the negative (−) residual for the three libraries ranged from 57% to 63%. Model 4 had a mean of 47% and the percentage of the negative (−) residuals for the three libraries ranged from 36% to 56%. Model 5 had a mean of 53% and the percentage of the negative (−) residuals for the three libraries ranged from 44% to 51%.

In addition, it appeared for each library studied that the model used made a difference in the percentage of negative (−) residuals for females and minorities for that library. In some cases, the differences for a library in the % of negative (−) residuals was again marked. In others, the differences were only slight.

In summary, the above findings on negative (−) residuals suggests, first, that the presence of salary disparities for both comparison groups studied (males against females and females and minorities against white males). Second, these findings suggest all models, regardless of the type of comparison groups, on average showed the percentage of disparities in salaries to be 47% or higher. Third, they suggest that comparing males against females instead of females and minorities may be preferable for two reasons. One, the comparison group, females and minorities, can only be used with libraries who have minorities reported. Two, comparing males to females keeps gender comparisons clear, while comparing females and minorities to white males does not.

However, the above analysis of findings only suggests there are salary disparities. In order to determine which model(s) is better for predicting salary disparities for academic librarians, an analysis of R-square values obtained using the STEPWISE method needs to be done.

EMPIRICAL ANALYSIS OF PREDICTOR MODELS

An analysis of the R^2 values and the strength of the predictor variables used in each of the five models tested is done to identify which of the five models tested was the strongest model for determining salary disparities.

The analysis of the strength of the predictor variables was based on R^2 values calculated for each predictor variable when doing the multivariable linear procedure using the method ENTER or the method STEPWISE. Predictor variables with higher R^2 values were considered to be stronger explanatory variables. If a variable was not included in the equation, a 0.50 limit had been reached which indicated the predictor variable was not considered an explanatory variable at all. Also, when using STEPWISE, the order in which the variables and their R^2 values were listed showed the extent to which each variable in the equation did or did not contribute to being a strong predictor variable. In STEPWISE, the first predictor variable listed (one) was the strongest predictor variable, the second predictor variable listed (two) was the next strongest predictor variable, etc.

Two comparison groups (males against females and females and minorities against white males) were studied. The findings for each model, by the comparison groups for each model, are discussed below.

Model 1/Method STEPWISE/3 Variables/ Males Against Females

For Model 1 in which the comparison group was males against females and the method STEPWISE was used, Table 3 shows the R^2 values obtained for the three predictor variables, highest degree level, year of birth, and year of highest degree level for all of the libraries being studied. For four of the five libraries, NIU, SIUC, UIC, and UIUC, the predictor variable, year of highest degree level, was the only predictor variable for which R^2 values were given. Also, the R^2 values were very similar to one another and ranged from 0.39287 to 0.40770. The other two predictor variables, highest degree level and year of birth, were not in the equation, and the 0.050 limits were reached. This means that, as predictor variables, they did not contribute to explaining salaries.

In contrast, for ISU the predictor variable year of birth had an R^2 value of 0.77324, and combined with the predictor variable highest degree level the R^2 value equaled 0.91152. Unlike NIU, SIU, UIC, and UIUC, where the year of highest degree level was the only predictor variable where R^2 values were given, for ISU this predictor variable was not in the equation and the 0.050 limits were reached.

Table 3. Model 1/Method STEPWISE/3 Variables/Males Against Females.

	Scott3		
	HDGLVL	YRBIRTH	YRHDLVL
ISU	2 R^2=0.91152 F=30.90499 Signif F=0.0007	1 R^2=0.77324 F=23.87017 Signif F=0.0018	X
NIU	X	X	1 R^2=0.39287 F=10.35364 Signif F=0.0054
SIUC	X	X	1 R^2=0.40770 F=5.50659 Signif F=0.0469
UIC	X	X	1 R^2=0.39699 F=9.21699 Signif F=0.0089
UIUC	X	X	1 R^2=0.40691 F=17.83856 Signif F=0.0003

X=variables not in the equation; 0.050 limits reached

Given the above, it does not appear that, in general, this model is a strong predictor of the disparity between women's salary levels and those of men in the librarian professorate in an academic library setting. For four of the five libraries studied, only one predictor variable of the three predictor variables in the equation contributed to explaining salaries. Even then, the R^2 values were low, ranging from 0.39287 to 0.40770.

Model 1/ Method STEPWISE/ 3 Variables/ Females and Minorities Against White Males

For Model 1, in which the comparison group is females and minorities against white males and STEPWISE is used, Table 4 shows the R^2 values obtained for

Table 4. Model 1/Method STEPWISE/3 Variables/Females and Minorities
Against White Males.

	Scott3		
	HDGLVL	YRBIRTH	YRHDLVL
NIU	X	X	1
			$R^2=0.35382$
			F=7.66595
			Signif F=0.0151
UIC	X	X	1
			$R^2=0.30104$
			F=5.59909
			Signif F=0.0342
UIUC	X	X	1
			$R^2=0.33385$
			F=11.52670
			Signif F=0.0025

X=variables not in equation; 0.050 limits reached.

the three predictor variables, highest degree level, year of birth, and year of highest degree level for three of the five libraries being studied. Only NIU, UIC, and UIUC had minorities represented in the data sets being studied. ISU and SIUC were not included in this part of the study because they did not report any minorities in their data sets.

For NIU, UIC, and UIUC the predictor variable year of highest degree level is the only predictor variable for which R^2 values were given. Also, the R^2 values were very similar to one another and ranged from 0.30104 to 0.35382. The other two predictor variables, highest degree level and year of birth, were not in the equation; 0.050 limits were reached. This means, as predictor variables, they did not contribute to explaining salaries.

In general, this model also does not appear to be a strong model for determining the disparity between women's salary levels and those of men in the librarian professorate in an academic library setting. First, this model is limited in its application in that it only can be utilized if there are minorities included in the data set. Second, for the three libraries studied, only one predictor variable of the three contributed to explaining salary disparities. Even then, the R^2 values were low, ranging from 0.30104 to 0.35382. Third, by having females and minorities compared against white males, gender dispersion is not clear.

Male librarians who are categorized as minorities would be included in the comparison group with females.

Model 2/ Method STEPWISE/6 Variables/ Males Against Females

For Model 2 in which the comparison group was males against females and STEPWISE was used, Table 5 shows the R^2 values obtained for the six predictor variables, highest degree level, year of birth, year of highest degree level, level of position, rank held, and years of previous library experience for all of the libraries being studied. Clearly, the strength of the six predictor variables and how much they contributed to explaining salaries for each individual library varied. For ISU, rank held was the only predictor variable where an R^2 value was given. This predictor variable alone explained 0.92076 of salaries. For NIU, three predictor variables, rank held, level of position, and year of birth, combined

Table 5. Model 2/Method STEPWISE/6 Variables/Males Against Females.

	Scott3 plus added predictors			Dowell	Kim	
	HDGLVL	YRBIRTH***	YRHDLVL	LVLPOSIT**	RANKHELD*	YRSPRLIEXP
ISU	X	X	X	X	1 R^2=0.92076 F=81.33546 Signif F=0.0000	X
NIU	X	3 R^2=0.85208 F=26.88137 Signif F=0.0000	X	2 R^2=0.78221 F=26.93738 Signif F=0.0000	1 R2=0.70325 R2=0.70325 Signif F=0.0000	X
SIUC	X	3 R^2=0.93206 F=27.43586 Signif F=0.0007	2 R2=0.82866 F=16.92762 Signif F=0.0021	X	1 R^2=0.53668 F=9.26673 Signif F=0.0160	X
UIC	X	X	X	1 R^2=0.66758 F=28.11549 Signif F=0.0001	2 R^2=0.86138 F=40.39061 Signif F=0.0000	X
UIUC	X	4 R^2=0.90104 F=52.35246 Signif F=0.0000	X	2 R^2=0.82508 F=58.96319 Signif F=0.0000	1 R^2=0.73569 F=72.36918 Signif F=0.0000	3 R^2=0.86264 F=50.24093 Signif F=0.0000

X=variables not in equation; 0.050 limits reached.
*Strongest predictor variable ** Second strongest predictor variable *** Third strongest predictor variable.

to explain 0.85208 of salaries. For SIUC, three predictor variables, rank held, year of highest degree level, and year of birth, combined to explain 0.93206 of salaries. For UIC, two predictor variables, year of highest degree level and year of birth, combined to explain 0.86138 of salaries. For UIUC, four predictor variables, rank held, level of position, years of previous library experience, and year of birth, combined to explain 0.90104 of salaries. When there was a larger sample, as was the case for UIUC, the regression holds more weight, possibly explaining why, for UIUC, years of library experience (three) and year of birth (four) were brought into the equation.

However, some general patterns appear to be present when looking at all the libraries collectively. For ISU, NIU, SIUC, and UICU the predictor variable rank held was a one in terms of order of predictor variables, making it appear as the strongest predictor of the six variables being tested. The next strongest predictor appears to be level of position. For NIU and UICU this variable was two in terms of its order of importance, and, for UIC, it was one. The third strongest predictor variable appears to be year of birth. The remaining three variables, highest degree level, year of highest degree level, and years of previous library experience, do not appear to contribute substantively to explaining salaries. The variable highest degree level is not included in the equation for any of the libraries as the 0.050 limits were reached. The variable year of highest degree level contributed to explaining salaries for only one library, SIUC. The variable years of previous library experience contributed to explaining salaries for only one library, UIUC.

Compared to Model 1, which used males against females as a comparison group, Model 2, which used males against females as a comparison group, appears to be a stronger model. Three variables, rank held, level of position, and year of birth, combined, to contribute to explaining salaries in Model 2. Also, in Model 2 the R^2 values, which ranged from 0.85208 to 0.93206, are much higher than those in Model 1. The higher the R^2 value, the more the predictor variables are contributing to explaining salaries. In Model 1 the comparison group males against females R^2 values ranged from 0.39287 to 0.40770 with the exception of ISU, and for females and minorities against white males the R^2 values ranged from 0.30104 to 0.35383.

Model 2/ Method STEPWISE/6 Variables/ Females and Minorities Against White Males

For Model 2 in which the comparison group was females and minorities against white males and STEPWISE was used, Table 6 shows the R^2 values obtained for the six predictor variables, highest degree level, year of birth, year of highest

Table 6. Model 2/Method STEPWISE/6 Variables/Females and Minorities Against White Males.

	Scott3 plus added predictors			Dowell	Kim	
	HDGLVL	YRBIRTH	YRHDLVL	LVLPOSIT**	RANKHELD*	YRSPRLIEXP
NIU	X	X	X	X	1 R^2=0.67347 F=28.87522 Signif F=0.0001	X
UIC	X	X	X	1 R^2=0.64453 F=23.57105 Signif F=0.0003	2 R^2=0.84692 F=33.19615 Signif F=0.0000	X
UIUC	X	4 R^2=0.87965 F=36.54681 Signif F=0.0000		2 R^2=0.79062 F=41.53658 Signif F=0.0000	1 R2=.68415 F=49.81824 Signif F=0.0000	3 R2=.83360 F=35.06857 Signif F=0.0000

X=variables not in equation; 0.050 limits reached.
* Strongest predictor variable ** Next strongest predictor variable.

degree level, level of position, rank held, and years of previous library experience, for three of the five libraries being studied. Only NIU, UIC, and UIUC had minorities represented in the data sets being studied. ISU and SIUC were not included in this part of the study because they did not have any minorities reported in their data sets. For NIU, rank held was the only predictor variable where an R^2 value was given. This predictor variable explained 0.67347 of salaries. For UIC, two predictor variables, rank held and level of position, combined to explain 0.84692 of salaries. For UIUC, four predictor variables, rank held, level of position, years of previous library experience, and year of birth, combined to explain 0.87965 of salaries. When there is a larger sample, as was the case for UIUC, more weight of the regression holds. This provides a possible explanation why, for UIUC, years of previous library experience (three) and year of birth (four) were brought into the equation.

Some overall patterns can be observed when looking at all the libraries collectively. The strongest predictor variable of the six predictor variables tested appeared to be rank held. Two of the three libraries, NIU and UIUC showed this predictor variable as a one in terms of order. UIC showed rank held as a two.

The next strongest predictor variable appeared to be level of position. UIC showed this variable to be a one, and UIUC showed this predictor variable as a two. The other three variables, highest degree level, year of birth and years of previous library experience, appeared not to contribute much to explaining salaries. The predictor variable highest degree level was not in the equation; 0.050 limits were reached. The other two predictor variables, year of birth (four) and years of previous library experience (three), only appeared in the equation for UIUC. This is likely the result of having a larger sample where the regression holds more weight.

This model appears to be a slightly stronger model than Model 1 in which the comparison groups are females and minorities against white males. In Model 2, the R^2 values were higher than those in Model 1 and ranged from 0.67347 to 0.87965. In Model 1, the R^2 values were much lower and ranged from 0.30104 to 0.35382. Also, as in Model 1 in which the comparison group was females and minorities against white males, this model was limited in its application in that it could only be utilized if there were minorities included in the data set. Also, as discussed earlier, gender dispersion became unclear when the comparison group is comprised of females and minorities and includes male minority librarians.

Model 3/Method STEPWISE/2 Variables/Males Against Females

For Model 3 in which the comparison group was males against females and the method STEPWISE was used, Table 7 shows the R^2 values obtained for the two variables, year of birth and year of highest degree level, for all of the libraries being studied. For four of the five libraries, NIU, SIUC, UIC, and UIUC, the variable year of highest degree level was the only predictor for which R^2 values were given. For ISU, the predictor year of birth was the only variable given.

Overall, this model, much like Model 1, does not appear to be a strong model for determining the disparity between women's salary levels and those of men in the librarian professorate in an academic library setting. The strongest of the two predictor variables, year of highest degree level, contributes to explaining salaries for four of the five libraries studied, but the R^2 values are low, ranging from 0.39297 to 0.40770.

Model 3/Method STEPWISE/2 Variables/ Females and Minorities Against White Males

For Model 3 in which the comparison group was females and minorities against white males and STEPWISE was used, Table 8 shows the R^2 values obtained for

Table 7. Model 3/Method STEPWISE/2 Variables/Males Against Females.

	Scott2	
	YRBIRTH	YRHDLVL*
ISU	R^2=0.77324 F=23.87017 Signif F=0.0018	X
NIU	X	R^2=0.39287 F=10.35364 Signif F=0.0054
SIUC	X	R^2=0.40770 F=5.50650 Signif F=0.0469
UIC	X	R^2=0.39699 F=9.21699 Signif F=0.0089
UIUC	X	R^2=0.40691 F=17.83856 Signif F=0.0003

X=variables not in the equation; 0.050 limits reached.
* Strongest predictor variable.

the two variables, year of birth and year of highest degree level, for three of the five libraries being studied. Only NIU, UIC, and UIUC had minorities represented in the data sets being studied. ISU and SIUC were not included in this part of the study because they did not report any minorities in their data sets.

For NIU, UIC, and UIUC the variable year of highest degree level was the only predictor for which R^2 values were given. Also, the R^2 values were very similar to one another and ranged from 0.30103 to 0.35382. For NIU, UIC, and UIUC, the variable year of birth was not in the equation.

Because two of the five libraries studied do not have minorities reported in their data sets, this model is limited in its application. Similar to the findings in Model 3 where the comparison group is males against females, only one of the two predictors in the equation contributed to explaining salaries. In addition, gender dispersion is unclear as the comparison group females and minorities includes male minority librarians. Also, the R^2 values were low and ranged from 0.30104 to 0.35382.

Table 8. Model 3/Method STEPWISE/2 Variables/Females and Minorities
Against White Males.

	Scott2	
	YRBIRTH	YRHDLVL*
NIU	X	R^2=0.35382 F=7.66595 Signif F=0.0151
UIC	X	R^2=0.30104 F=5.59909 Signif F=0.0342
UIUC	X	R^2=0.33385 F=11.52670 Signif F=0.0025

X=variables not in the equation; 0.050 limits reached.
*Strongest predictor variable.

Of the three models discussed above, it appears first that Model 1 and Model 3 tend to be weak models in comparison with Model 2. Second, models that use the comparison group males against females are stronger models than models that use the comparison group females and minorities against white males for two reasons.

Model 4/Method STEPWISE/5 Variables/Males Against Females

For Model 4 in which the comparison group was males against females and STEPWISE was used, Table 9 shows the R^2 values obtained for the five predictor variables, year of birth, year of highest degree level, years of previous library experience, level of position, and rank held for all five of the libraries being studied. For three of the five libraries, ISU, SIUC, and UIUC, the variable rank held was the only predictor variable for which R^2 values were given. For UIC, the only variable for which an R^2 value was given was the level of position. For NIU, two predicators, rank held and level of position, were in the equation. For all five libraries, the variables year of birth, year of highest degree level, and years of previous library experience were not in the equation in that 0.050 limits were reached. R^2 values ranged from 0.53668 to 0.92076.

Table 9.   Model 4/Method STEPWISE/5 Variables/Males Against Females.

	YRBIRTH	YRHDGLVL	YRPRLIEX	LVLPOSIT	RANKHELD*
			Scott2 plus added Predictors		
ISU	X	X	X	X	R²=0.92076 F=81.33546 Signif F=0.0000
NIU	X	X	X	R2=0.78221 F=26.93738 Signif F=0.0000	R2=0.70325 F=37.91692 Signif F=0.0000
SIUC	X	X	X	X	R²=0.53668 F=9.26673 Signif F=0.0160
UIC	X	X	X	R²=0.66758 F=28.11549 Signif F=0.0001	X
UIUC	X	X	X	X	R²=0.73569 F=72.36918 Signif F=0.0000

X=variables not in the equation; 0.050 limits reached.
* Strongest predictor variable.

In general, it appears, when looking at the range of R^2 values of all the models with the comparison group males against females discussed to this point, this model (0.53669 to 0.92076) is similar to Model 2 (0.85208 to 0.93206), and Models 1 (0.39287 to 0.40770) and 3 (0.39699 to 0.40770) are similar to each other. Also, Model 2 and Model 4 appear based on their higher R^2 values to be stronger predictor models than Model 1 or Model 3.

Model 4/Method STEPWISE/5 Variables/Females and Minorities Against White Males

For Model 4 in which the comparison group was females and minorities against white males and the method STEPWISE was used, Table 10 shows the R^2 values obtained for the five variables, year of birth, year of highest degree level, years of previous library experience, level of position, and rank held, for three of the five libraries being studied. Only NIU, UIC, and UIUC had minorities

Table 10.   Model 4/Method STEPWISE/5 Variables/Females and Minorities
Against White Males.

	Scott2 plus added predictors				
	YRBIRTH	YRHDGLVL	YRPRLIEX	LVLPOSIT	RANKHELD
NIU	X	X	X	X	R^2=0.67347 F=28.87522 Signif F=0.0001
UIC	X	X	X	1 R^2=0.64453 F=23.57105 Signif F=0.0003	X
UIUC	X	X	X	2 R^2=0.79062 F=41.53658 Signif F=0.0000	1 R^2=0.68415 F=49.81824 Signif F=0.0000

represented in the data sets being studied. ISU and SIUC were not included in this part of the study because they did not report any minorities.

Both NIU and UIUC ranked the variable rank held as the strongest predictor for these two libraries. In contrast, for UIC, level of position was the only variable in which an R^2 value was given. For all three libraries the predictor variables year of birth, year of highest degree level, and years of previous library experience were not in the equation because the 0.050 limits were reached.

In general, it appears when looking at the range of R^2 values of all models where the comparison group is females and minorities against white males discussed to this point, this model (0.64453 to 0.79062) is similar to Model 2 (0.67347 to 0.87965), and Model 1 (0.30104 to 0.35382) and Model 3 (0.30104 to 0.33385) are similar to one another. Also, Model 2 and Model 4 appear, based on their higher R^2 values, to be stronger predictor models than Model 1 or Model 3.

Model 5/Method ENTER/3 Variables/Males Against Females

For Model 5 in which the comparison group was males against females and the method ENTER was used, Table 11 shows the R^2 values obtained for the three stepwise/strongest predictors, rank held, level of position, and year of birth, for all of the five libraries being studied. With the exception of SIUC,

Table 11.   Model 5/Method ENTER/3 Variables/Males Against Females.

	Stepwise/ Strongest Predictors
	R^2 values
ISU	0.95571 F=35.96015 Signif F=0.0008
NIU	0.83634 F=20.44036 Signif F=0.0001
SIUC	0.67613 F=4.17523 Signif F=0.0646
UIC	0.86561 F=25.76340 Signif F=0.0000
UIUC	0.84072 F=42.22718 Signif F=0.0000
Mean	0.83490

the R^2 values were very similar to one another and ranged from 0.83490 to 0.95571. On average, the R^2 value was 0.83490.

Overall, this model appeared to be the most efficient model for predicting salary disparities. This model was a combination of the strongest predictor variables which were determined using the method STEPWISE on Models 1–4 and only three predictor variables were needed to do calculations. In contrast, the other two models (Model 2 and Model 4), which on average had slightly higher R^2 values, required more predictor variables to do calculations. Model 2 required data gathering on 6 predictor variables and Model 4 required data gathering on 5 predictor variables.

Model 5/Method ENTER/3 Variables/Females and Minorities Against White Males

For Model 5 in which the comparison group was females and minorities against white males and the method ENTER was used, Table 12 shows the R^2 values obtained for the three stepwise/strongest predictors, rank held, level of position, and year of birth for three of the five libraries studied. The R^2 values were very similar to one another and ranged from 0.80749 to 0.85208. On average, the R^2 value was 0.84017.

Overall, this model in comparison with the other models which compared females and minorities against white males appeared to be the most efficient model for predicting salary inequities. This model was a combination of the strongest predictor variables which were determined using the method STEP-WISE on Models 1–4 and only three predictor variables were needed to do calculations. In contrast, the other two models (Model 2 and Model 4), which on average had slightly higher R^2 values, required more predictor variables to do calculations. Model 2 required data gathering on 6 predictor variables and Model 4 required data gathering on 5 predictor variables.

As was the case for all models studied in which the comparison group was females and minorities against white males, two of the five libraries studied did not have minorities reported in their data sets, resulting in this model being limited in its application. Also, this again resulted in gender dispersion being unclear as it included some males.

Table 12.   Model 5/Method ENTER/3 Variables/Females and Minorities
Against White Males.

	STEPWISE/ Strongest Predictors
	R^2 values
NIU	0.85208
	F=26.88137
	Signif F=0.0000
UIC	0.86096
	F=22.70419
	Signif F=0.0001
UIUC	0.80749
	F=29.36166
	Signif F=0.0000
Mean	0.84017

Overall Summary of Findings/Comparison of Means of R^2 Values
for All Models Using the Method ENTER

The R^2 values for all the models studied in which the comparison groups were males against females and the method ENTER was used were compared with each other in Table 13. Certain patterns emerged which provided indications of which model or models were stronger models for predicting salary discrepancies based on an analysis of the means of their R^2 values. Model 1 had an R^2 value mean of 0.57312 and Model 3 had an R^2 value mean of 0.50626. Compared to the R^2 value mean of the other models studied, these R^2 values were considered low. Model 1 and Model 3 appeared to be weak models for predicting salary disparities. Model 2 had an R^2 value mean of 0.92341 and Model 4 had an R^2 value mean of 0.92118. These R^2 value means were considered high. Model 2 and Model 4 appeared to be equally strong models for predicting salary discrepancies. Model 5 had an R^2 value mean of 0.83805. Model 5, while it did not have as high an R^2 value mean as Model 2 or Model 4, was still considered high. Of the three models with high R^2 value means, Model 5 appeared to be the most efficient model as it, for data gathering, required the least number of predictor variables.

The R^2 values for all the models studied in which the comparison group were females and minorities against white males and the method ENTER was used were compared with each other in Table 14. Similar patterns to those discussed in Table 13 provided indications of which model or models were stronger models for predicting salary discrepancies based on an analysis of the means of their R^2 values. Model 1 had an R^2 value mean of 0.37921 and Model 3 had an R^2 value mean of 0.35238. Compared to the R^2 value means of the other models studied, these R^2 values means were considered low. Model 1 and Model 3 appeared to be weak models for predicting salary discrepancies. Model 2 had an R^2 value mean of 0.87931 and Model 4 had an R^2 value means of 0.87848. There R^2 value means were considered high. Model 2 and Model 4 appeared equally strong models for predicting salary discrepancies. Model 5 had an R^2 value mean of 0.83484. Model 5, while it did not have as high an R^2 value mean as Model 2 or Model 4, is still considered high.

Of the three models with high R^2 value means, Model 5 appeared to be the most efficient model as it required, for data gathering, the least number of predictor variables.

As stated previously, when using the comparison group in which females and minorities were compared against white males, two constraints occurred. First, the model is limited in its application in that it can only be utilized if minorities are included in the data being reported. Only three of the five

Table 13. R^2 Values/Males/Females/Method ENTER.

	Model 1 3 predictor variables (Scott3)	Model 2 6 predictor variables (Scott3 plus added predictors)	Model 3 2 predictor variables (Scott2)	Model 4 5 predictor variables (Scott2 plus added predictors)	Model 5 3 predictor variables (Stepwise/ predictor)
ISU	0.91152	0.98511	0.84391	0.98320	0.95571
	F=17.16981	F=22.05496	F=16.2190	F=35.11476	F=35.96015
	Signif F=0.0046	Signif F=0.0440	Signif F=0.0038	Signif F=0.0073	Signif F=0.0008
NIU	0.47783	0.86965	0.41857	0.86960	0.85208
	F=4.27042	F=12.23172	F=5.39923	F=16.00422	F=20.44036
	Signif F=0.0245	Signif F=0.0003	Signif F=0.0171	Signif F=0.0001	Signif F=0.0001
SIUC	0.62476	0.95709	0.42877	0.94452	0.67613
	F=3.32991	F=11.15244	F=2.62709	F=13.62026	F=4.17523
	Signif F=0.0979	Signif F=0.0369	Signif F=0.1409	Signif F=0.0127	Signif F=0.0646
UIC	0.41604	0.90417	0.40460	0.90254	0.86561
	F=2.84977	F=14.15330	F=4.41695	F=18.52048	F=25.76340
	Signif F=0.0820	Signif F=0.0004	Signif F=0.0344	Signif F=0.0001	Signif F=0.0000
*UIUC	0.43543	0.90104	0.43543	0.90104	0.84072
	F=9.64086	F=52.35246	F=9.64086	F=52.35246	F=42.22718
	Signif F=0.0008	Signif F=0.0000	Signif F=0.0008	Signif F=0.0000	Signif F=0.0000
MEANS	0.57312	0.92341	0.50626	0.92018	0.83805

*For UIUC in Model 1, the variable YRBIRTH was not in the equation, and in Model 2, the variables YRHDGLVL and HDGLVL were not in the equation; tolerance = 1.00E-04 limits reached.

libraries studied met this condition. Second, gender dispersion become unclear as minorities included males who were minorities.

In summary, given the above findings, Model 1 and Model 3 appeared to be weak models for predicting salary disparities for the library faculty groups studied. Model 2 and Model 4 appeared to be equally strong models for predicting salary disparities and Model 5 appeared to be the most efficient model. In addition, it appeared that using the comparison groups that compared males against females eliminated problematic constraints associated with using the comparison groups that compared females and minorities against white males. In this comparison, gender dispersion is clear. However, when the

Table 14. R^2 Values/White Males/Females and Minorities/Method ENTER.

	Model 1 3 predictor variables (Scott3)	Model 2 6 predictor variables (Scott3 plus added predictors)	Model 3 2 predictor variables (Scott2)	Model 4 5 predictor variables (Scott2 plus added predictors)	Model 5 3 predictor variables (Stepwise/ predictor)
NIU	0.45176 F=3.29609 Signif F=0.0579	0.85914 F=9.14904 Signif F=0.0021	0.38099 F=4.00065 Signif F=0.0443	0.85800 F=12.08416 Signif F=0.0006	0.83634 F=26.88137 Signif F=0.0000
UIC	0.32246 F=1.74506 Signif F=0.2156	0.89915 F=11.88813 Signif F=0.0013	0.31272 F=2.73010 Signif F=0.1054	0.89778 F=15.80937 Signif F=0.0003	0.86096 F=22.70419 Signif F=0.0001
UIUC	0.36342 F=6.27977 Signif F=0.0070	0.87965 F=36.54681 Signif F=0.0000	0.36342 F=6.27977 Signif F=0.0070	0.87965 F=36.54681 Signif F=0.0000	0.80749 F=29.36166 Signif F=0.0000
MEANS	0.37921	0.87931	0.35238	0.87848	0.83484

Note: No minorities are reported in the data sets for ISU and SIUC. Therefore, only NIU, UIC and UIUC who reported having minorities are tested using the White Males/Females and Minorities/Stepwise/5 variables model.

comparison group used is white males compared to females and minorities, gender dispersion is less clear because there are males in the females and minorities comparison group. Second, in this study two of the five libraries did not have any minorities in their librarian professorate, limiting the applicability of this comparison.

Other Observations on Findings

Of the five libraries studied, both ISU (*n*=26) and SIUC (*n*=25) had a small number of cases (n) in comparison with the other three libraries. Because ISU and SIUC had a small number of cases (*n*) causing R^2 values to go up, caution should be taken when interpreting their findings, because with the above caution in mind and based on the findings, it appeared a critically low mass of n might have been reached for ISU. For all models in which the comparison group was males against females, the R^2 values for ISU ranged from 0.84391 to 0.98320. However, for SIUC, which also had a low number of cases, the R^2 values were not consistently high and ranged from 0.42877 to 0.95709. It appeared, when looking at the SIUC patterns of R^2 values, they were more consistent with the other libraries studied, other than ISU. The above may in part be explained because of another difference between ISU and the other libraries studied. The Scott (1977) study recommended that at least 15 white males be included in the group being studied. ISU reported only 9 white males. SIUC reported 10 white males, and the three remaining libraries all had the recommended 15 or above white males. A possible explanation for the differences between ISU and SIUC, both of which have a small number of cases (n), is that there may be problems with R^2 values when there are low sample sizes combined with a single-digit number of white males (or males when the comparison group is males against females). In effect, for ISU, the combination of the presence of a low number of cases (n) and a low number of white males (males) appeared to result in R^2 values being high.

CONCLUSIONS

Based on the findings of the negative ($-$) residuals analysis, the conclusions below were drawn. First, the presence of negative ($-$) residuals at all of the libraries studied, regardless of the model used or comparison group used, provided evidence that gender-based wage disparities existed within the librarian professorate that was studied. This conclusion was consistent with the research reported in the literature on the extent and magnitude of wage-gap problems for librarians in general and more specifically for academic librarians. The literature indicated that, in general, librarians experience gender-based wage-gap problems. More specifically in the literature, it was indicated that academic libraries experienced gender-based wage-gap problems to a greater extent than other types of libraries. Second, the magnitude or extent of the wage disparities appears to pose an equity problem for all the libraries studied. Regardless of the model or the comparison group used, on average, the percentage of disparities in salaries was 47% or higher. Third, in developing a recommended methodology for determining gender-based wage disparities for the librarian professorate, a new problem associated with gender-based wage disparities was uncovered. Gender-based wage discrimination persisted in the librarian professorate studied in spite of the fact that the professional group was predominately female and the women in the groups studied participated, as other members of the group, in making recommendations on salary increases. This finding suggested that the dynamics of gender-based wage discrimination are much more complex and subtle than has been previously represented in the research literature. It is suggested that, in female-dominated professions, multiple layers of gender-based wage discrimination exist. Gender-based wage disparities may not only exist for the group as a whole, as suggested in some of the theories, but also within the group itself. Fourth, while the negative (−) residual analysis suggested the existence and extent of gender-based wage disparities, as a tool of analysis it was limited in its scope as a tool of analysis. This type of analysis was not able to contribute to identifying which of the five models tested was the strongest model for determining salary disparities. Another type of analysis was needed.

Based on the findings of the empirical analysis of the five predictor models, the following conclusions were drawn. First, for the five models tested, some models appeared to provide stronger predictors of salary disparities than other models in determining salary disparities for the librarian professorate. Model 1 and Model 3 appeared to be weak predictor models. Both these models only used predictors that were used in the Scott (1977) study. Model 2 and Model 4 appeared to be stronger predictor models, but Model 5 appeared to be the most efficient predictor model. All these models included added variables. Second, it was found that which predictor variables were used made a difference in determining the strength of the model's ability to predict librarian professorate salary disparities. However, some variables that were used to predict salary disparities for other faculty groups resulted in weak predictor models, Model 1 and Model 3, when applied to the librarian professorate.

In doing both of the analyses above, some additional observations on the findings were noted and additional conclusions drawn. One observation was that, if the comparison group which was used was white males against females and minorities, it appeared to be problematic in two distinct ways. First, gender dispersion was unclear. Second, a problem was created when a group being studied did not report any minorities. In effect, the comparison was no longer white males compared against females and minorities, but males against females. Given the above, it was concluded that the comparison group of males against females was preferable.

What conclusions could be drawn regarding the three hypotheses that were discussed earlier in this chapter? The first hypothesis stated that the recommended methodology for demonstrating and measuring gender-based wage discrimination in the Scott study, i.e. predicted regression, that is applicable for "regular faculty" is applicable without any modification for the librarian professorate. It was concluded that while the procedures used in the Scott (1977) study were applicable, two modifications were necessary. First, different predictor variables needed to be used in the regression equation. Second, the comparison group needed to be males compared with females instead of white males compared against females and minorities.

The second hypothesis stated that the precision of Scott's model is significantly strengthened by including additional predictors. Based on the findings, it was concluded that the predictor variables that were used to determine gender-based wage disparities for the "regular faculty" appeared not to be strong predictors for the librarian professorate. Model 1 and Model 3, which used only predictor variables used in the Scott (1977) study, were the weakest models of the five models tested. Additional predictor variables needed to be used in combination with one of the variables used in the Scott model, year of birth. The three variables, year of birth, rank held, and level of position, were stronger predictors in determining gender-based wage disparities for the librarian professorate.

The third hypothesis was that the precision of Scott's model is strengthened by comparing the salaries of males against females instead of comparing males with females and minorities. Based on the additional observations on the findings it was concluded that the precision of Scott's model was indeed strengthened by limiting this comparison in this way.

Based on the conclusions discussed above it appears to be possible to create a recommended methodology for determining the disparity between women's salary levels and those of men in the librarian professorate. Of the five models tested, Model 5 appeared to be the most efficient and is recommended for use to determine and measure gender-based wage discrimination for the librarian professorate. The same multiple linear regression procedures which were used in the Scott model could be used in the librarian professorate model. However, there were two basic differences in the models. First, the predictor variables used were different. Second, the comparison group used was different. However, the librarian professorate model accomplished the same objectives as the Scott model. It used a limited number of easily obtainable predictor variables and was able to flag women for whom there is apparent salary inequity using a statistical technique that is widely in use. Any library could easily gather the data and do the procedures cost effectively.

IMPLICATIONS AND CONTRIBUTIONS OF THE STUDY

The findings of this study have the potential for contributing to the research literature that was reviewed. First, the findings of this study provide recent, empirical evidence of the continuing existence of a wage-gap problem for the librarian professorate. While there have been numerous studies of librarians' salaries indicating gender differences, the number of empirically based studies on academic librarians has been limited. The findings in this study can help fill this critical gap in the library research literature.

But the most significant contributions of this study's findings may be in the contributions it makes towards influencing future research directions in the literature on theoretical frameworks, where scholars continue to debate and attempt to explain reasons for gender-based wage disparities. The problem uncovered here, that gender-based wage disparities exist even in a female-dominated profession where the females, along with others in the group, participate in setting salary increases, has significant potential for triggering new thinking on why this happens and for sharing these new ideas through the development of new theoretical frameworks which deal with some of the problems and ideas set forth in this study. Few, if any, studies currently address the problem uncovered here. These findings should contribute to the initiation of follow up studies being done and reported in the literature.

Also, there is a serious gap in the literature of librarianship dealing with the use of multiple regression techniques as a method for measuring wage disparities. This study's findings being reported in the literature would provide the field of librarianship, not only with a model, but also a better understanding of how multiple regression techniques can be used to determine, measure, and explain gender-based wage disparities.

RECOMMENDATIONS FOR FURTHER STUDY

While it provides a way to determine and measure gender-based wage disparities of individuals who are part of a group in a specific library, who are subject to the same personnel criteria, the methodology recommended by this study is not likely to totally correct salary discrepancies for female-dominated professions. The findings suggest that finding a way to determine and measure wage disparities between individuals within a group is only the first step. There must also be a way to determine and measure gender-based wage disparities that exist between female-dominated groups and non-female-dominated groups to gender-based wage disparities are to be totally corrected. As discussed earlier, the findings uncovered a new research problem which needed to be explored and explained – the fact was that gender-based wage disparities persisted in the librarian professorate studied in spite of the fact that the professional group was female dominated and that the women in the groups studied participated, as did other members of the group, in making recommendations on salary increases. Based on these findings, it was suggested that gender-based wage discrimination may be more complex than previously suggested in the research literature. It was also suggested that there may be multiple layers of gender-based wage discrimination occurring within female-dominated professions. In effect, the causation of gender-based wage discrimination may be more complex than the research literature to date has identified. The librarian professorate is considered a female-dominated occupational group. In effect, the suggested "recommended methodology" used for the librarian professorate in this study may only be peeling away one layer of gender-based wage discrimination, the layer that deals with gender-based discrimination among individuals within a group. It does not address gender-based wage discrimination that may be occurring between groups as a result of occupational segregation. The idea that for female-dominated professions, such as the librarian professorate, there may be multiple layers of gender-based wage discrimination occurring needs to be the subject of further research.

Also, this recommended methodology was developed based on data from one type of academic library, academic libraries that were public and state-supported. Another question posed for further research is, does this model hold for other types of academic libraries? Can the same recommended methodology be used for the librarian professorate who work in private academic libraries or in community college academic libraries? If the answer to this question is yes, then the recommended methodology as developed in this study can perhaps be used as a general model of any type of academic library. Can there be, as is the case with the Scott model, a general recommended methodology to determine and measure gender-based wage disparities for the librarian professorate?

REFERENCES

Annual report of the economic status of the profession, 1995–96. (1996, March-April). *Academe*, *82*(2), 14–108.

Ashraf, J. (1996). The influence of gender on faculty salaries in the United States. *Applied Economics*, *28*, 857–864.

Association of Research Libraries (1998). *ARL annual salary survey, 1997–98*. Washington, D.C.: Association of Research Libraries.

Bazemore v. Friday 41 FEP Cases 92 (1986).

Becker, G. S. (1971). *The economics of discrimination*. (2nd ed.). Chicago: University of Chicago Press.

Becker, E., & Lindsay, C. M. (1995, January). Male/female disparity in starting pay. *Southern Economic Journal*, *61*, 628–643.

Bergmann, B. R. (1975, October). How to analyze the fairness of faculty women's salaries on your own campus. *AAUP Bulletin*, *61*(3), 262–265.

Bergmann, B. R. (1986). *The economic emergence of women*. New York: Basic Books.

Blau, F. D., & Kahn, L. M. (1994, May). Rising wage inequality and the U.S. gender gap. *The American Economic Review*, *84*, 23–28.

Blaug, M., & Sturges, P. (Eds) (1983). *Who's Who in Economics: A Biographical Dictionary of Major Economists 1700–1981*. Brighton, England: Wheatsheaf Books Ltd.

Blinder, A. S. (1973, Fall). Wage discrimination: Reduced form and structural analysis. *Journal of Human Resources*, *8*, 436–455.

Blum, L. M. (1991). *Between feminism and labor; The significance of the comparable worth movement*. Berkeley: University of California Press.

Carson, C. H. (1996, October 15). Beginner's luck: A growing job market. *Library Journal, 121*, 29–30+.

Corcoran, M. E. (1979). Work experience, labor force withdrawals, and women's wages: Empirical results using the 1976 Panel of Income Dynamics. In: C. B. Lloyd, E. S. Andrews & C. L. Gilroy (Eds), *Women in the Labor Market* (pp. 216–245). New York: Columbia University Press.

Dean, J. M., & Clifton, R. A. (1994). An evaluation of pay equity reports at five Canadian universities. *Canadian Journal of Higher Education, 24*(3), 87–114.

DeNavas, C. (1997). *Money income in the United States, 1996: with separate data on valuation of noncash benefits*. Washington, D.C.: U.S. Dept. of Commerce, Economics, and Statistics Administration, Bureau of Census.

Dowell, D. R. (1986). The relation of salary to sex in a female dominated profession: Librarians employed at research universities in the South Atlantic census region. (Doctoral dissertation, University of North Carolina at Chapel Hill, 1986). *Dissertation Abstracts International, 48*(2), 238.

Dowell, D. R. (1988, May). Sex and salary in a female dominated profession (multiple regression analysis of library salaries). *The Journal of Academic Librarianship, 14*, 92–98.

England, P. (1982). The failure of human capital theory to explain occupational sex segregation. *Journal of Human Resources, 17*(3), 358–370.

England, P. (1992). *Comparable worth: Theories and evidence*. New York: Aldine de Gruyter.

England, P., Chassie, M., & McCormick, L. (1982). Skill demands and earnings in female and male occupations. *Sociology and Social Research, 66*(2), 147–168.

England, P., Farkas, G., Kilbourne, B. S., & Dou, T. (1988). Explaining occupational sex segregation and wages: Findings from a model with fixed effects. *American Sociological Review, 53*, 544–558.

Feldberg, R. (1980). Union fever: Organizing among clerical workers, 1900–1930. *Radical America, 14*(3), 53–67.

Female workers win salary dispute. (1998, July 30). *Chicago Tribune*. section 1, p.14

Ferber, M. A., & Kordick, B. (1978). Sex differentials in the earnings of Ph.D.s. *Industrial and Labor Relations Review, 31*, 227–238.

Fuchs, V. R. (1988). *Women's quest for economic equality*. Cambridge: Harvard University Press.

Fudge, J., & McDermott, P. (1991). *Just wages; A feminist assessment of pay equity*. Toronto: University of Toronto Press.

Genova, B. K. L., Gill, K., & Cole, E. (1977). *A study of salary determinants within the SUNY Librarians's Association between 1973 and 1974*. (ERIC Document Reproduction Service No. ED134 189)

Gray, M. W., & Scott, E. (1980, May). A statistical remedy for statistically identified discrimination. *Academe*, 174–181.

Harris, R. M., Monk, S., & Austin, J. (1986, June). M. L. S. graduates survey: Sex differences in prestige and salary found. *Canadian Library Journal, 43*, 149–153.

Hildenbrand, S. (1997, March 1). Still not equal: Closing the library gender gap. *Library Journal, 122*(4), 44–46.

Hill, M. A., & Killingsworth, M. R. (Eds) (1989). *Comparable worth: Analysis and evidence*. Ithaca, N. Y.: ILR Press

Hutner, F. C. (1986). *Equal pay for comparable worth ; The working woman's issue of the eighties*. New York: Praeger.

Johnson, G., & Stafford, F. (1974). The earnings and promotion of women faculty. *American Economic Review, 64*, 888–903.

Jones, K. F. (1987, October 15). Sex, salaries, & library support. *Library Journal, 112*, 35–41.

Katz, D. (1973). Faculty salaries, promotions, and productivity at a large university. *American Economic Review, 63*, 469–477.

Kieft, R. N. (1974, April). Are your salaries 'equal'?: A techniques for assessing the equability of your salaries. *College Management, 9*(23), 23.

Kim, U. C. (1980). *A statistical study of factors affecting salaries of academic librarians at medium-sized state-supported universities in five Midwestern states.* (Doctoral dissertation, Indiana University, 1980).

Kyrillidou, M., & Maxwell, K. A. (Eds) (1995). *ARL Annual Salary Survey 1995-1996.* Washington, DC: Association of Research Libraries.

Kuper, A., & Kuper, J. (1996). *The social science encyclopedia.* (2nd ed.). London: Routledge.

Loeb, J., & Ferber, M. (1971). Sex as predictive of salary and status on a university faculty. *Journal of educational measurement, 8*, 235–244.

Looker, E. D. (1993). Gender issues in university: the university as employer of academic and nonacademic women and men. *Canadian Journal of Higher Education, 23*(2), 19–43.

Mellor, E. F. (1984, June). Investigating the differences of weekly earnings of women and men. *Monthly Labor Review, 107*, 9–15.

Milkman, R., & Townsley, E. (1994). Gender and the economy. In: N. J. Smelser & R.Swedberg (Eds), *The Handbook of Economic Sociology.* (pp. 600–619). Princeton: Princeton University Press; New York: Russell Sage Foundation.

Moore, M. V., & Abraham, Y. T. (1995, Fall). Comparable Worth: Is it a moot issue? Part III controversy, implications, and measurement. *Public Personnel Management, 24*(3), 291–313.

Moore, N. (1993). *Evaluation of faculty salary equity models.* Unpublished doctoral dissertation, Arizona State University.

National Committee on Pay Equity. (1998). *About the National Committee on Pay Equity* (NCPE). [On-line] Available: http://www.feminist.com/fairpay.htm 7/15/98 9:32AM

Norusis, M. J. (1993). *SPSS for windows base system users's guide release 6.0.* Chicago, IL: SPSS, Inc.

Pay equity: issues and strategies. (1987). Chicago, IL: American Library Association, Office for Library Personnel Resources.

Paul, E. F. (1989). *Equity and gender: The comparable worth debate.* New Brunswick, N.J.: Transaction.

Pezzullo, T. R., & Brittingham, B. E. (Eds) (1979). *Salary equity: Detecting sex bias in salaries among college and university professors.* Lexington, MA: Lexington Books.

Polachek, S. (1981). Occupational self-selection: A human capital approach to sex differences in occupational structure. *Review of Economics and Statistics, 63*, 60–69.

Ray, J. M., & Rubin, A. B. (1987, January). Pay equity for women in academic libraries: An analysis of ARL salary surveys., 1976/77–1983/84. *College & Research Libraries, 48*, 36–49.

Raymond, R. D., Sesnowitz, M. L., & Williams, D. R. (1990). The contribution of regression analysis to the elimination of gender-based wage discrimination in academia: A simulation. *Economics of Education Review, 9*(3), 197–207.

Rutherford, D. (1992). *Dictionary of economics.* London: Routledge.

Rytina, N. F. (1982, April). Tenure as a factor in the male-female earnings gap. *Monthly Labor Review, 105*, 32–34.

St. Lifer, E. (1994, November 1). Are you happy in your job? LJ's exclusive report. *Library Journal, 119*, 44–49.

Methodology for Determining Gender Salary Inequity 173

Sayer, L. (Ed. & compiler) (1996). *SLA biennial salary survey 1996.* Washington, D.C.: Special Libraries Association.

Scott, E. L. (1977). *Higher education salary evaluation kit: a recommended method for flagging women and minority persons for whom there is apparent salary inequity and a comparison of results and costs of several suggested methods.* Washington, D.C.: American Association of University Professors.

Scott, R. L., Farr, W. K., Flanders, E. L., & Spiers, B. E. (1993, Fall). Professional and salary characteristics of librarians employed by senior colleges of the University System of Georgia. *Southeastern Librarian, 43,* 46–51.

Sobel v. Yeshiva University 32 FEP Cases 150 (1977).

Sorensen, E. (1989). The wage effects of occupational sex composition: A review and new findings. In: M. A. Hill & M. R. Lillingsworth (Eds.), *Comparable Worth: Analysis and Evidence.* (pp. 57–79). Ithaca, New York: Cornell University.

Sorensen, E. (1991). Exploring the reasons behind the narrowing gender gap in earnings. (*Urban Institute Report* 91–92). Washington, D.C.: Urban Institute Press of America

Sorensen, E. (1994). *Comparable worth; Is it a worthy policy?* Princeton: Princeton University Press.

Sowell, T. (1984). *Civil rights: Rhetoric or reality?* NY: W. Morrow.

Spaulding v. University of Washington 740 F.2nd. 686 (1984).

SPSS 6.1 for windows. (Release 6.1.3). [Computer software]. (1995, December 5). Chicago,IL: SPSS Inc.

Stevenson, M. H. (1975). Relative wages and sex segregation by occupation. In: C.B. Lloyd. (Ed.), *Sex, Discrimination, and the Division of Labor.* (pp. 175–200). New York: Columbia University Press.

Titus, E. (1997). 1996 Illinois academic library statistics: summary tables and graphs compiled from the 1996 Integrated Post-Secondary Education Data Systems (IPEDS). Springfield: Illinois State Library.

Wage-gap. (1996). http://www.feminist.com/wagegap.htm 7/15/98 9:39 AM Willborn, S. L. (1986). A comparable worth primer. Lexington, Mass.: Lexington Books.

Zellner, H. (1975). The determinants of occupational segregation. In: C. Lloyd (Ed.), *Sex, Discrimination, and the Division of Labor* (pp. 125–145). New York: Columbia University Press.

THE EFFECTS OF AUTOMATION ON HIRING PRACTICES AND STAFF ALLOCATION IN ACADEMIC LIBRARIES IN TENNESSEE

Murle E. Kenerson

INTRODUCTION

Academic libraries and librarianship are experiencing a profound change as a result of automation of its holdings as well as its networking capabilities to the world of information resources. Part of this change can be traced to several causes, e.g. budget restrictions and a more diverse student population requiring an ever-widening array of services from the university library and its staff. But the most significant shifts in academic library procedures relate directly to the introduction of new information and communication technology. Any change brings, as a matter of course, problems to be resolved, the need for both strategic and tactical planning, and an affirmative response to a rapidly evolving environment. The approaches ultimately selected must provide a vehicle through which extensive substantive, organizational, and technological change may be accomplished. It is to be acknowledged that technologies can provide alternatives and divergencies not yet fully understood. Further, as stated by Convoy (1982), "change is no longer a one-time event to be, at one time, completed . . . Changes of role, function, people, and technology will provide a rolling environment, an exacting, emerging, and exhaustive place to be"

Advances in Library Administration and Organization, Volume 18, pages 175–223.
2001 by Elsevier Science Ltd.
ISBN: 0-7623-0718-8

(p. 95). As a result, the murky, dynamic environment it creates challenges the planner who seeks to meld the desired outcomes, qualities, and direction into a format that can provide a creative, effective beginning to the process of change.

Automation has brought extensive change to all institutions that deliver information, including the academic library. Yet little has been reported in the literature on the impact of automation on library personnel and, most critically, on the management of staffing practices in the Information Age. Dakashinamurti (1985) found that attention has veered to consideration of the human factor in machine-human interface, and as a consequence "not much data has been gathered about this aspect in libraries, particularly on the effects of technology on library staff and staff management" (p. 343). As Olsgaard (1989) noted, "staffing is literally the alpha and the omega of all automation projects but probably receives the least attention" (p. 484).

Administrators and researchers alike have tended to focus on the process of automation itself rather than on how automation affects traditional staffing patterns, arrangements, and/or reassignments. Hill (1988) pointed out that no single formula can apply to all academic libraries in regard to hiring procedures or staff allocation. However, "major variables can be identified . . . that will ensure the maximum potential of library personnel is realized through the interaction between the various aspects of the tasks, the support equipment, and the people performing the tasks" (pp. 89–92). Managing every facet of this interaction becomes crucial in that the quality of service is rapidly emerging as the vital criterion in the Information Age.

The decisions and choices afforded to the academic library administrator in today's technology environment have seldom been easy or well-defined, especially those relating to the role of automation in hiring and staffing arrangements. If, for example, service is to become the standard by which the academic library is judged, will automation advance or retard performance? According to Dougherty (1994), the technological transformation of the academic library may bode an end to the deep commitment to service on which the profession is dependent or offer unparalleled opportunities. "One can build a logical case for either point of view" (p. 355). It may be asserted, as do DeKlerk and Euster (1989), that changes resulting from the use of technology is a solution to library problems. Technology has permitted staff shifts, the reassignment of positions, with new or retrained staff, changes in job designs and descriptions and, in some instances, the hiring of more paraprofessionals to fill particular slots. These authors believed that this "organizational metamorphosis helps to cope with growth in programs and added responsibilities in such areas as public services, collection development, and systems management" (p. 459). Still others, including DeBruijn (1986) perceive library automation, not as an opportunity to improve service delivery, but as a means of reducing labor costs through changes in the number or type of staff hired in a period of frozen budgets. If new services could not be added, at least none would have to be cut.

Problems arising from the interplay between automation and the traditional hierarchical reporting lines, long a hallmark of the academic library organization, cannot readily be discounted. Various workplaces have experienced a democratization process with the introduction of technology. Organizational and procedural changes occur most notably in decision making and responsibility levels. Woodsworth, Allen, Hoadley, Lester, Molholt, Nitecki and Wetherbee (1989, p. 135) held that administration in the academic library has assumed a very different configuration as a result of automation. Past management patterns that stress supervision, control, and the enforcement of standards have been replaced by those emphasizing a more collegial workplace. The role of the administrator has become one of facilitating, coordinating, and orchestrating the work of the staff. Decision making has been at the lowest possible action level. Automation has thus standardized and diffused responsibilities among staff and administration. However, participatory decision making and increased responsibility levels assumed staff had the skills to respond to new demands for accountability and service delivery.

The question has been whether the academic library that is rapidly automating should look for (or could find) those skills within the ranks of professionally degreed librarians or seek a more diversified staff under the assumption that automation creates new jobs, requiring new skills, and new training. Should new employees be hired with needed technological skills or should existing staff be retrained to meet new roles and expectations? Would more para-professionals with lesser salary requirements provide more benefit to library patrons than professional librarians or would service diminish? What consideration should be given to the employment of individuals with degrees in areas other than librarianship; how might they be best allocated within the academic library? Should expertise with computers and other technical operations be the primary skill sought or are there other skills of import in the mix? These and similar questions relating to staffing arrangements within the academic library pose very real problems for the administrator in the Information Age. The problem to be examined, therefore, is the effect that automation has on staffing arrangements and other personnel shifts within the academic library.

Purpose of the Study

The primary purpose of this study is to examine in depth the effects of automation on hiring practices and staff allocations in public and private libraries in Tennessee. It investigates how automation has changed the ways in which libraries function, significantly altering job roles and descriptions and work patterns. It focuses on the idea that automation is more than the sum parts of mechanizing those functions once done by manual, labor intensive efforts. Technological conversion is a complex, complicated, and on-going process that has influence and will continue to influence who is hired, how professionals and paraprofessional are recruited, what skills are essential in the various levels of employment, and how staff may be allocated. Although the responsibility for ensuring a smooth transition to automation resides with the director or administrator, few models have been available to assist in planning. Few guidelines have been provided to help administrators transition to automation in the automated academic library setting. At the same time, the degree to which oganizations have managed hiring and staff allocation has determined the degree to which libraries have maintained quality service as those services shift and change as a result of automation. This study will provide insight into how high standards of service might be retained and even heightened with the proper utilization of human and machine resource management.

Significance of the Problem

Automation in the academic library has presented administrators with both challenge and opportunities. Increasingly diverse student populations and expanding programs and course offerings have created demand for access to information in all formats, packaging that information to suit the user's unique needs. At the same time, budget constraints contribute to the pressure surrounding the hiring and allocation of the people needed to produce quality service in both traditional and nontraditional configurations.

Although most academic library administrators are sure of the benefits of automation and recognize its potential to effect a fundamental transformation in library services, there is little agreement about the impact of technology on hiring practices, training and retraining staff, and the allocation of personnel. In one study (Daniels, 1995), over half of academic library administrators believed that automation had not affected staff selection; the majority still placed a higher value on the traditional skills of librarianship than on understanding automated systems. However, the bulk of the respondents felt that computer-literate staff enhance service delivery to patrons. Corbin (1991), stated that "every staff member, regardless of their responsibilities or relationship to the automated system, should have a basic understanding of computers and automation" (p. 39). But beyond this rather narrow perception of needs emanating from the realities of the automated work place, data on which one might construct a model for staffing and positioning personnel is limited.

This must change as Dakashinamurti (1985), put it now that the Information Age is upon us. He contended that "It is imperative that we use technology wisely and most effectively by ensuring that the maximum potential of library personnel is realized through productive management techniques" (p. 351). An obvious implication is that there is a need for perceptive, organized leadership. Planning for the changes wrought by automation is not a neutral process; the development and selection of alternatives impose the constraints of value judgments, the identification of priorities, the anticipation of issues and priorities, the anticipation of issues and problems, and the definition and clarification of jobs, skills, and accountability. The formation and utilization of models, (that is, conceptual schemes, combined with knowledge about the impact of automation on the mission of the academic library) can facilitate change.

This study was driven by the absence of viable planning tools for use by those charged with staffing the academic library as automation spreads to all facets of library work. More resources for those overseeing the transition to the new technologies, especially the building of personnel competencies and components for change, become of prime importance. The introduction of technology creates new, potentially serious problems, particularly if all factors in the work system are not adequately considered. Few guidelines have yet been established to assist in this highly complex, often contradictory, process. Certainly, a single solution to hiring and allocating staff in the automated academic library cannot be articulated. But, it is necessary to develop more precise data by which to base improved responses to the questions surrounding the adjustments and reorganization of positions in this unique work setting.

As noted above, the scope of the research was concerned with public and private academic libraries in Tennessee, and study findings and conclusions related only to these academic libraries. Therefore, the extrapolation of study results to other different situations and circumstances should be done, if at all, with a degree of caution.

Research Questions

The following questions were posed in order to guide the study analysis of hiring and staff allocation practices as these policies pertain to Tennessee academic libraries, both public and private.

(1) Is there a significant difference in the numbers of professional librarians employed before and after an academic library automates?
(2) Is there a significant difference in the numbers of support staff employed before and after an academic library automates?
(3) When an academic library automates, is there a significant difference in the numbers of "other" professionals hired (for example, computer specialists and computer programmers)?
(4) Do job descriptions, content, and status change when an academic library automates?

Assumptions of the Study

This study was based on the following assumptions:

(1) Automation will significantly alter the hiring policies of academic libraries in terms of numbers in the various categories, skills sought, and in recruitment methods.
(2) Automation will significantly alter the allocation of staff in the academic library due to changes in job content, responsibilities, skills, levels, and service demands.
(3) While the desire is to determine the effects of automation on hiring and allocation of staff in academic libraries, it is not realistic to expect that all changes can be identified.
(4) It is assumed that the respondents taking part in this study will provide accurate and reliable information in their assessment of the effects of automation on hiring and staff allocations in the academic library.

THE LITERATURE

It is a truism that conventional academic library services can only provide a partial answer to the needs of higher education as the 20th century draws to a close. It has become evident to most of those who administer them that a fundamental shift to new technology is occurring and will continue to expand our capability to locate and retrieve information from all sources. This has been evident in public policy discussions and legislation at the national level, producing the National Information Infrastructure Act of 1993 (Love, 1993). Other examples include models for regulating federal policy, proposed in H.R. 3459, the Improvement of Information Access Act, which affects how information has been disseminated over the Internet (Love, 1993; Mitchell & Saunders, 1993).

Arnold, Collier, and Ramsden (1993) found that, although the technology for this transition is already available, "many practical and theoretical factors are inhibiting the actual transformation of the academic library" (p. 3). Oberg (1995) explained that more widespread utilization and acceptance of automation has been hampered by the same difficult issues within the library profession for most of the past hundred years and which remain mostly unresolved. These include "self utilization, role definition and articulation, task overlap, educational requirements, certification, and status" (p. 3). Over the past 20 years, the growth of automation and declining budgets have intensified the need to address in a structured way how the problems associated with the automated academic library can be best managed. Smalley (1994) believed that library administrators have not sufficiently considered the tradeoffs imposed by the new technologies and have not been "proactive enough in shaping their design and function" (pp. 360–361). Going to the heart of the problem, McCrank (1986) found that it was the technology itself that too frequently takes the spotlight, not the more difficult problems of methodology, management, professionalism, and "the all too human element" (p. 63). Transition demands the thoughtful management of people, not just an adequate budget and the ideal machinery to bring the academic library into the electronic age successfully. Manson (1987) stated that technology and the field of library automation have never been static and have become increasingly less so in the recent past. One benefit of this revolution is that it forces a re-evaluation of the management and tasks of the organization in order that sensible decisions can be made. Recognizing that the decisions and choices afforded to the library administrator are seldom easy or well-defined, Manson further noted that "an awareness of the factors involved in the change process, in particular, the effects of automation on hiring practices and staffing arrangements, have been largely ignored in studies of the 'new' academic library" (pp. 108–109). Until these questions are explored in depth, the problems presented by the technological environment impacting on personnel management will remain. For as Dakashinamurti (1985) noted, "technology only provides tools . . . yet it is the complete tasks to be performed which requires consideration, the interaction between the various aspects of the tasks, the support system, and the people needed to carry out those tasks" (p. 343).

Emerging Issues of Automation

In 1992, Oberg warned that librarianship will not attain full professional status until librarians come to terms with the staffing dilemmas associated with the changes caused by automation. He suggested that the roles of librarians as well as support staff must be redefined and clarified. While some hold that librarianship will have to be essentially "reinvented" in order to resolve such issues as the role of support staff and paraprofessionals vis-à-vis the professional librarian, the challenging new tasks created by automation, and the demand for quality service, Pritchard (1995) stated that new roles and tasks should have been derived from the old. These include:

(1) relating user needs to information availability;
(2) managing complex technological, financial, and bureaucratic systems;
(3) designing interconnected technical systems, organizational structures, and human interfaces;
(4) selecting and organizing information resources;
(5) teaching and consulting; articulating logical and intuitive insights about information;
(6) interacting with the external environment through the formation and articulation of information policy (p. 3–4).

Still, as Veaner (1994) pointed out, automation has transformed most library workers into "knowledge workers" and the once simple distinction between librarians and support staff fails to describe adequately the complexities of today's workplace. Many support staff, and some librarians argue that the Masters in Library Science (MLS) should not constitute a barrier to advancement, and that librarianship out to be competency, rather than degree based. But Veaner warned that "control of the library's programs and fiscal responsibility are inherent in the librarian's position, and, by definition, cannot be delegated" (p. 390).

On the other hand, Schnelling (1992) believed that noting that change is occuring is an understatement. He contends that "it would be more appropriate to call this process an overturning of administrative procedures, management structures, working tools, and financial frameworks" (p. 324). Choices have been made as a result of these alterations. Automation requires training and oftentimes the upgrading of posts. This has led, in turn, to a leveling out of the traditional hierarchical order sustained by different qualifications, training, and type of duties. "Eventually, everyone will be expected and prepared to do the same work in the fully automated library" (p. 339). Handling these problems, while maintaining the divide between librarians and other staff will be impossible, and it is safe to admit this leveling is inevitable and to "support it in terms of enriching individual staff members" (p. 339). Needless to say, this also entails the adjustment of salaries to compensate the change in the level of the responsibilities of the staff.

Malinconico (1994) addressed the "deskilling of labor and fragmentation of jobs" (p. S2) brought about by the use of new technologies in the library setting. Automation once introduced into information handling activities meant that work took on, to some extent, the characteristics of mass production jobs: the "mechanical pacing of work, repetitiveness, minimum skill requirements, pre-determined use of tools and techniques, and surface mental attention." At the same time, productivity and service quality improve if work organizations are developed which encourage experimentation and provide opportunity for interpersonal interaction. "The models and methods used in the past to adopt and apply new technology are distinctly inappropriate when used for the application of electronic technologies to information handling activities" (S. 4).

Perplexity of Funding Automation
It was the position of Robinson and Robinson (1994) that library administrators are at present confronted with a number of hard choices. "They have the difficult task of reallocating existing dollars in a effort to introduce new services in response to customer demands, to capitalize on technological advances, and to ensure that the staff operations needed to produce products and services are in place and supported" (p. 421). It was stated that it has been easy to manage competing interests when money was available; the problem has been to "run an effective operation when money was in short supply" (p. 421). For as Thompson (1992, p. 2) attested, automation does not save money; it improves services. Adding to the difficulties is the component mentioned by McCrank (1992). Academic libraries find that once introduced "automation spawns demands for increased services; it has a spiraling effect" (p. 90). After automating, academic libraries typically experience dramatic increases in more extensive and intensive usage among patrons. This has lead to pressure on the library to increase staff and to introduce more technology – neither of which is easy in a era of fiscal restraint – or to more efficiently allocate staff, whether in positioning, retraining in computer skills, or "in change of style and methods of delivering service" (p. 91).

Horney (1987) also discussed the issue of whether or not automation saved money. It was noted that the question must be viewed primarily in terms of staffing. "The extent to which automation leads to savings in staff time or enables a less expensive staffing configuration is a matter of debate." While staff positions usually have not reduced in the wake of automation in the academic library, there was a reluctance to eliminate positions that were already budgeted for. "It has also been observed that new automation-related tasks often occupy time saved by the elimination of steps in the previous manual system." And, because "automation often costs significant amounts for equipment and operations, considerable savings in salaries may be needed to balance the new expenses." Given the complexity of the issues and problems associated with automation, few academic libraries have been cognizant of all the extended effects and results of technology on hiring and staff allocations at the point that automation was introduced.

Hiring Practices Associated with the Automated Academic Library

Staffing Cost Considerations

One of the most profound influences on the academic library have been the high expectations on the part of library users for the various electronic services. According to Martin (1994) "most academic libraries have moved so far down the path of electronic dependence for technical operations and reference that there is no way of going back" (p. 479). However, unless payment is made for the electronic services demanded by librarians and patrons, alike, they cease. These external services now consume a much larger part of the library budget than formerly, reducing the degree of flexibility available to administrators. Libraries have had "to rethink their budgets for staffing as well as the organizational environment in which that staff will function. The same forces changing the library materials budget in light of automation are also at work in the personnel budget" (p. 487). Martin (1991) noted that, although it had not generally been thought of in such a way, the hiring of personnel was the purchase of time and expertise, and it has been appropriate to ensure that both was used as beneficially as possible.

Goudy (1993) found that the transfer to nonprofessional personnel has been overcoming one of the inherent weaknesses of many academic libraries, over reliance on professional staff for many activities which could be conducted by staff with less training. These include a number of tasks once understood to be professional in nature: copy cataloging, acquisitions, interlibrary loan, and circulation. In large measure, these alterations have been based on the increasing availability of electronic work tools. The switch was fostered by the need to stretch the staff budget by using lower grades of personnel where feasible. Indeed, Goudy discerned the absolute reduction in total numbers of staff members and budgets. In fact, personnel and general expenditure shares of the budget in libraries have declined precipitously in the decades of the 1970s through the 90s.

Shifting Roles of Staffing the Academic Library

DeKlerk and Euster (1989) believed that the shortage of librarians in some areas has contributed to both the shifting of roles within the library and the nature of hiring practices. "Many library school graduates have acquired information skills that can be put to use in other job areas so that fewer graduates are choosing traditional library positions in academic libraries" (p. 465). This shortage has required the administrator to seek out alternatives and/or supplements for the professional staffing arrangements. At the same time, new positions have emerged which must be filled. Microcomputer information specialist, systems librarian, and coordinator of database search services are examples of recent computer-related job titles. The proliferation of technology-driven positions in the academic library is one reason for the findings by Craghill (1989) that "no substantial net reductions in staffing levels have resulted, or appear likely to result from the implementation of automation" (p. 19).

The effect of automation on the selection of new staff was in debate, however. In a study by Dover (1991, pp. 78–89), 52.5% of the respondent's said that automation had not affected staff selection, while 40% thought that it had. However, there were several signs that suggested this position was shifting. Although the majority (60%) still placed a higher value on traditional library skills, a number of libraries (16%) looked for automation skills on an equal level, and 9% even placed experience of automation as the most important attribute. Johnson (1991) found that nearly all respondents (98.1%) to her study felt some minimal understanding of automated systems was necessary to be considered for employment. Becker (1985) discerned that staffing the academic library was a difficult task. In her survey, the most important requirements for the automated workplace were good communication skills, business knowledge, and a personable and supportive personality. However, while these attributes were desirable, she concluded that they "must be combined with technical proficiency" (p. 17). Research by Palmini (1994, p. 122) on academic libraries in Wisconsin, determined that, according to 47% of her respondents, some computer background was necessary for support staff positions. Only 13% stated that no computer background would be necessary. Some felt that applicants' willingness to learn, was a more important component than actual knowledge.

Hill (1988) pointed out that the traditional distinction between professional and nonprofessional staff "is that professionals make judgments, while nonprofessionals follow rules" (p. 92). This divide had implications in the hiring process. However, as bibliographic networks and other external databases expand, and as "automation imposes standardization on an increasing number of technical services operations, the body of work over which judgment must be exercised is shrinking." As technological developments enable increased speed and efficiency in professional tasks, the number of staff needing professional supervision decreases. In his case study of the impact of automation on staffing at a medium-sized academic library, Kraske (1978) noted that, as the direct result of technology, six professional positions were eliminated over a ten-year period, while the number of clerical and other staff positions increased over that same decade from 32 to 45. The qualifications for many of these support positions were, at the same time, raised in accordance with the added responsibilities. Thus positions once labeled Typist Clerk I and II, calling for general office skills, became Library Technician I and II. Employees sought for these jobs were expected to possess a working knowledge of the library's automated systems and some familiarity with cataloging rules.

Selecting Qualified Applicants for the Academic Library
Birdsall (1991) acknowledged that academic libraries have been in competition with each other in filling professional positions. The recruitment of the professional librarian has been an important and complex undertaking. He provided a series of steps as a guide for finding the best candidates available, arguing that a national search for the right person is required rather than simply promoting someone already in the organization. Birdsall also stipulated that a search committee offer the proper means of conducting the process. Among the steps advocated by Birdsall are: establishing a time frame for the hiring procedures, developing selection criteria, drafting and distributing the announcement of the job description, statements on qualifications, salary, benefits, and deadlines for applications. It was recommended that a "weighted scoring instrument, based on qualification criteria, be developed as an aid in the evaluation of applications" (p. 279).

Balbach (1989) advocated that librarians take some hiring risks, avoiding "safe hires in which the person looks the same as others in the unit." She warned about placing too much importance on impressions formed in the interview. What should be stressed is the contribution the individual might make to the strength of the staff and its overall effectiveness. As Birdsall concluded, "continued success in the recruitment of librarians may depend on a skillful marketing of the library and its environment, and a substantial investment of staff time and library monies" (p. 238). An interesting example of "risk-taking in staffing" the academic library's professionals was detailed by Bonta (1990). As he described it, the Pennsylvania State University Library, when staffing for CD-ROM service, opted not to hire a professional librarian for the assistant position, but selected instead a person completing a doctorate in archaeology. However, the individual had "considerable experience as a user of libraries and computers; even more importantly, she had a strong sense of service and a willingness to learn" (p. 9).

Montague (1993) reported that automation's impact on library staffing requirements has been extremely variable. Some academic libraries have increased staff, some have decreased staff, others have remained essentially the same. Some have been able to provide more services and process more materials without adding to staff levels; others have been unsuccessful in this area, either having to add personnel or cut back in service. Montague also asserted that, in general, staffing has remained constant as the academic library changes to automation. While different types of workers might be hired, there is usually a "one-to-one" force in operation in replacement of staff, meaning that job titles and content are very different, but numbers are not. Still, it is maintained that the challenge remains to select and hire those who understand, or can be trained to understand, automated systems, the role of technology in the library's mission, and their own role in making automation work.

According to a study by Heim and Moen (1990) encompassing 3,484 students at Louisiana State University School of Library and Information Science, that challenge of finding the well-trained, qualified professional for the academic library was less than what was supposed. Slightly over 35% of the candidates for a masters' degree in Library Science indicated that they preferred a first position in an academic library over work in other settings. This serves to demonstrate that, although it is a "seller's market" to some extent, the academic library is in a favorable position to recruit top candidates if hiring practices are not cumbersome, and if a deliberate and well-fashioned recruitment and selection process has been put in place.

Toward this end, McDaniels and Schmidt (1989) discussed in some depth the utilization of computers in implementing personnel selection. They described two computer systems which proved to be valuable in the identification of capable applicants and their probable job performance levels. It was discovered that the "validity of personnel selection procedures varied with the cognitive complexity of jobs" (p. 78). It was also found that methods focusing on credentials were least useful, while those procedures focusing on job-related achievements and accomplishments were most effective. Nor was length of experience necessarily a useful criterion; increased job experience was linearly related to job performance only for approximately five years. After five years, the amount of job experience no longer differentiated among those employees with varying levels of productivity.

In examining the results of the research, Auld (1987) noted that there was a real problem associated with finding librarians for technical services. A survey discerned curricular shortcomings which Auld suspected may lead to lengthy on-the-job training even for otherwise highly-qualified, newly-hired personnel. Technical courses, Auld stated, tended to have been too general and theoretical as well as too short for the intended purpose. Auld's study suggested that the problem may well stem from the belief that automation spells the end to the need for many technical services staff members, particularly in such areas as cataloging and acquisitions. But as Saunders (1996) noted, a problem exists as well in the blurring of lines between information technologists and librarians. Although the two have different academic preparation and credentials, they have overlapping functions in meeting users' needs. Hence, the employer seeking staff and operating on a lean budget may be limited to a choice between specialties.

Koenig (1993) held that the technology-driven dimension has changed what the library employer looks for in new hires. The "essence of that change is that the employer now looks for someone not only with technical and professional skills but also with managerial skills . . ." "A dramatic burgeoning of interest in information management . . . is indicated by the number of business schools which are initiating such programs . . . in effect, merging library and information science with that of business management courses" (p. 284). Haywood (1991) asserted that library schools are being required by the changes in employment opportunities both to serve the traditional community and to have served as a special purpose business school if its graduates are to be hired by today's academic library. Koenig (1993) explained that what was once regarded as a service profession must now, because of technology and smaller budgets, be converted into part of the market economy. The employer, in consequence, no longer searches for the candidate with the knowledge of how to operate information systems, but for those who also know how to create such systems. At the very least, they must be able to choose among various software packages and build an information system based upon one of them; "it is increasingly the case that academic librarians create information systems rather than just use them" (pp. 280–281).

Several rather specific concepts on what academic libraries most desire in the context of new hires were spelled out by Woodsworth, Allen, Hoadley et al. (1989). These capabilities and skills included: knowledge of cognitive and disciplinary research processes and of psychology; technological sophistication; well-developed interpersonal skills; knowledge of information policy development and analysis; and planning skills. In addition, "librarians will be assertive risk-takers and synthesizers and have the ability to function in an atmosphere of ambiguity and change" (pp. 134–135). On the whole, academic librarians are better educated, and all personnel, whatever the level of employment for which they are hired, must be willing to assume more responsibility and to function as part of a "team" effort. DeKlerk and Euster (1989) added that any clerical or paraprofessional position opened in most automated academic libraries now demand the ability to use microcomputers and training in searching database. Employment policies also spoke of new technologies. For example, one library uses "multiple patterns in defining new jobs; unique combinations of tasks and responsibilities. Another is looking at job rotation and yet another, the training of patrons." DeKlerk and Euster (p. 466) commented as well on the new and different skills administrators of academic libraries have been seeking. These include: skills in budgeting, strategic planning, educational technology, and time management. Increased emphasis has been placed on financial accountability in higher education and on personal characteristics such as flexibility, ability to deal with change and ambiguity, and willingness to work as part of a team.

While hiring practices have tended toward incremental change in meeting new staffing needs in the academic library, it has become readily apparent that a number of newly created positions have emerged from the automation process. Many are held by professionals who are not librarians but have been drawn from other disciplines. Others have been paraprofessionals, in part, handling duties once performed only by a degreed librarian. Technical positions, directly related to automation, for example, microcomputer information specialist and coordinator of database search services were required and have demanded the development and implementation of a unique selection methodology not known in even the recent past.

The staffer selected may lack the technical knowledge necessary for decision making, but may possess the desired technical skills to meet the demand for expertise in library computer applications. Increasingly, the librarian with the traditional Masters of Library Science, is now required to possess new skills brought on by the Information Age. As DeKlerk and Euster (1989) reported, "it appears that in response to the growing complexity of providing library services in academe . . . new strategies for staffing academic libraries to meet the sophisticated demands of users in an online environment are needed" (p. 467). The rapid rate of change and the need to adapt old structures and adopt new ones will continue to complicate hiring procedures in the discernible future.

Allocation of Staff in the Automated Academic Library

Strategic Planning for Staffing the Academic Library
Robinson and Robinson (1994) asserted that "managing in a time of change requires that library directors think strategically and challenge assumptions about their mission, their staff, and their clientele" (pp. 420–421). In a period of shrinking financial resources, the control of costs and the maintenance of service quality assumed primary importance. Automation has improved productivity, yet has not usually resulted in major savings. However, as de Bruijn (1986) pointed out, automation provided more flexibility in how the administrator allocates the job duties of librarians and support staff. Martin (1989) noted that the academic library must identify the appropriate ways to deliver new information services.

Martin found that libraries have tended to incorporate the new technology into their buildings and operating procedures in a "haphazard" way. An example cited was the introduction of CD-ROM as a tool in the reference area, "with little fanfare and no changes in staffing arrangements" (p. 376). On the other hand, many perceive the advent of information technologies as a golden opportunity to entirely restructure the work environment. And often, new organizational structures within the library are the only way to cope successfully with change.

De Gennaro (1994) stated that any alterations in such procedures as staffing arrangements have been highly dependent upon whether the library takes a "direct approach to automation or adapts a more evolutionary course" (p. S10). If the "direct" route is pursued, the working theory revolves around the library being a "single complex operating system" (p. S10); all its varied operations are interrelated and interconnected. It is only logical then that it be treated as a unit. Tasks might be designed and implemented as a series of modules, but all must be designed as part of a whole. By way of contrast, the evolutionary approach envisioned the library moving from a traditional manual system to a complex, automated one over a period of time and in a more "piecemeal" fashion. This was, and remains, the most common approach to academic library automation and as such has implications for allocation of staff. Martin also discerned the two approaches as falling on a continuum, with most academic libraries remaining on the conservative end. However, the policy is utilized remains dependent upon such elements as the nature of the institution, the characteristics of the library staff, the preferences of administrators, responses of clientele, timing, and the availability of resources, to name a few factors.

Dyer, Fossey and McKee (1993) agreed that whether the approach to automation was revolutionary or evolutionary in design, its introduction will produce change in the working environment and affect staff structure reflected in skill requirements, responsibility, status and career paths – as well as in patterns of relationships and communications. While the nature of technology may preclude some options in staff allocation, "it should be recognized that decisions on the best form of work organization were likely to be based upon historical practice, attitudes towards the workforce, and factors sometimes quite separate from the nature of the technology itself" (p. 3). Horney (1987) pointed out that, with automation, certain jobs may no longer be significant. "Others may combine, or become interdependent, due to the way in which functions interrelate within the system" (p. 69). Decentralization can become possible once information previously located with individuals becomes available across a network of computer terminals.

Prince and Burton (1988) concluded after studying the impact of technology on the structures of three academic libraries, there was no evidence of a simplistic cause and effect mechanism in operation. "Rather, it is necessary to examine the interactions of the entire socio-technical system" (p. 80). Managerial policies and preferences have had a major role in staffing arrangements. They found that automation affects jobs both directly and indirectly. They referred to a study which reported that twice as many more skilled jobs were created than lost in the insurance industry, "because the introduction of information technology enabled tasks not previously performed manually to take place and made time for more creative work" (p. 77). The authors discovered as well that academic-related staff in libraries were being replaced by para-professionals as the clerical work was being enhanced and posts upgraded. Professional positions declined as technology simplified task, making work simpler to perform and thus suitable for support staff. It would appear that even within a single research effort, disagreement was evident in how staff is being rearranged when automation was being introduced.

Kiesler, Obrosky and Pratto (1987) held a rather different perspective, arguing that automation brought about a "reskilling of both professional and clerical grades. This was the case particularly where management style matched task needs and staff abilities without concern over job titles" (p. 148). New professional posts such as systems librarians were being created even as others were lost to clerical and para-professional workers. It was stated that a decrease in the level of supervision by professional staff was another result of technology, while there was an increase in the supervision of support staff and students by library administration. It was the contention of DeKlerk and Euster (1989) that, as a result of automation, there have been fewer professional workers in technical services because of changes to the cataloging process and, at the same time, an increased demand on the public service area where professional librarians now spend more of their time. The authors contend that this has changed the orientation of the library "from a warehouse to a client-centered approach" (p. 468). Dyer, Fossey, and McKee (1993) were of the opinion that information technology has been in itself neutral as an instrument of change and that libraries vary considerably in the effect that automation has had on their overall management structure.

Case Studies of Staff Allocations in Academic Libraries
In a case study examining the general effects of automation in a medium-sized academic library, Kraske (1978) found that technology has resulted in overall savings in labor costs brought about by staff realignment, primarily by a decrease in professional positions and an increase in the numbers of support staff. The new allocation of staff is most notable in the creation of two new units, notably systems development and operations, which were said to have

had a centralizing impact upon the library. It was pointed out that it is difficult to establish a direct relationship between automation and the reduction or transfer of positions since the connection is not always well defined. The major organizational change in this library involved the transfer of seven positions to data processing from cataloging. Several professional positions in acquisitions, serials, and in cataloging were also eliminated due to automation.

Atkinson and Stenstrom (1984) also perceived no clear-cut indication that positions were lost as a result of automation. A restructuring of the academic library through automation does not magically bring savings. Library managers must have the rationale and vision, that their staff buys into, that creates a sound mix of automation and personnel. The information age has forced automation into libraries and managers must arrange staffing patterns to accommodate the technology. Cain (1986) thought that, while most academic libraries have not been involved in major reorganizations of staff with the implementation of automation, some cataloging departments have essentially been disbanded and replaced with a processing unit, composed of nonprofessionals. The professional catalogers have then been dispersed to different public service units.

> A librarian's duties now will be organized by scholarly discipline rather than to function, that is, each professional will provide an array of technical and public services, from book selection and cataloging to reference and database searching, with particular responsibility for a narrowly defined set of academic subjects (p. 183).

It was also noted that the principal effect of online bibliographic services has been to reduce the need for local cataloging. This has permitted administrators to shift personnel from cataloging to other library work. But, Berman (1978) took the opposite view, contending that automation results in the need for more professional cataloguers to improve mass production cataloguing and to create a catalog at the local level which makes the collection more accessible to users. Name and authority work still must be done and substandard cataloging data available from a cooperatively developed database must be brought up to standard. Perhaps it may be, as Cline and Sinnott (1983) stipulated, that "as more ways are found to exploit the capabilities of new technology, job specifications of staff will keep changing" (p. 157), as will the allocation of that staff.

DeBruijn (1986) described differences in the allocation of staff in the University of British Columbia Library following automation. As discovered in other studies, the cataloguing functions were most subject to reorganization. In this area, the administrative hierarchy was flattened, and the three original divisions were transformed into two. The heads of these divisions reported directly to the Assistant University Librarian for Technical Services. The Catalogue Records Division assumed most of the responsibilities of the former

Original Cataloguing Divisions, with the exception of added copies and volumes processing. "Librarians and support staff were reorganized into subject or language units with each unit headed by an original cataloguer" (p. 18). The Catalogue Products Division took over the responsibilities of the former Catalogue Preparations Division as well as added copies and volumes processing. De Bruijn said that, as automation continued, librarians and senior library assistants came to make up a larger proportion of the staff and student assistants and junior library assistants form a small proportion. At the same time, the ratio of librarians to support staff increased. This increase was due, not to an increase in the numbers of librarians employed, but to a large decrease in the number of support staff. De Bruijn recognized that this finding contradicted much of the research in the area, but maintained the validity of his conclusions.

While not discussing the automated library specifically, Langdon (1995) offered insight into how technology relates to staff organization. Management specialists, it was pointed out, argued that firms can no longer afford to keep large staffs through thick and thin. Instead, because of budget constraints, whether in the public or private sector, they are being encouraged to retain a smaller core staff, augmenting it as necessary with temporary employees, consultants, specialists and others not on the permanent payroll. As long-term demand for certain services becomes clearer, entities may commit themselves to permanent employees in particular specialties; with so many institutions now in a state of rapid change, it was postulated that "outside contracting for filling certain slots is the practical solution for short-term needs" (p. 76). Bartlett (1995) highlighted the issues that "downsizing" often creates. He noted that there must remain a sufficient number of employees so that no one feels overwhelmed. Employees must be matched to the tasks they are expected to perform. Because change frequently mandates training and retraining, the learning style of each employee has to be considered.

A large-scale study of 24 academic libraries by Veaner (1993) discerned a number of problems inherent in the management of the computer facility and its personnel. One factor was the difficulty in ascertaining who was really in charge of the facility. It was stated that the organization and management of library automation activities demonstrate development phases. For example, in some larger academic libraries, the staff of the computer center may be half or even more than half that of the library. The expansion of both components produces labor intensive pressures. Library processing staff doubles and triples while more sophisticated technology needs to be supported by greater numbers of still more highly qualified systems programmers, communication experts, and user services staff. Differences in service capacity, resource utilization, and staff allocations thus represent a key political issue affecting academic libraries. It was held also that too much of the library staff was still centered in "house-keeping and internal technical processing," its format is still too rigid; "its general fixity of organization and structure remains at variance with changing patrons expectations and interests" (p. 51).

Library Management and Staffing Procedure

Managing Automation and Personnel Cost in the Academic Library

Stuart-Stubbs (1994) stated, only semi-comically, that the administrator's job in automation is people: "to have the right people in the right jobs, to be on the right side of all the right people, and to know the right thing to do at the right time" (p. S16). If libraries are to survive as viable institutions, managers must solve an array of problems associated with staffing, putting the right people in the right places, as Stuart-Stubbs would have it. A number of researchers including Dyer, Fossey and McKee (1993) believed that the impact of new technology on the organization depends on the prevailing managerial climate, because automation in and of itself has not predetermined positive or negative effects on staffing arrangements. Rather, its impact "seems to depend on the motives underlying management decisions to introduce the technology in the first place" (p. 11). Information technology is, in itself, neutral as an instrument of change and libraries vary in the effect that it has on their overall structure. McLean (1981) found that libraries have, in the main, adopted automated systems for two basic reasons: (1) to be more efficient in what is already being done, and (2) to offer services and support which could not be achieved manually. The relentless pressure of labor costs was the primary spur forcing libraries to turn to the computer. An essential objective for management then was the utilization of technology as to reduce labor cost.

More recently, library managers have justified the installation of automated systems by citing such factors as improved service to library patrons and reduced processing time. Boss (1984) felt that "improved services are a more compelling reason to automate than are possible reductions in costs" (p. 7) and that there was little evidence that labor costs were in fact reduced by technology. A comparison of an inefficient manual system with an efficient automated system may show substantial savings in terms of staff costs, but the role played by automation in reducing those costs was obscured. And while Kraske (1978) found that the "overall effect of automation has been a savings in labor costs in the academic library" (p. 13), his study did not include cost figures or compar-isons. On the other hand, Hegarty (1985) felt that labor costs can be reduced through automation, and that the resulting savings can be used to support other services or to meet external demand for budget cuts. However, Veaner (1994) held that libraries often failed to realize significant staff savings through automation because, "as computers and systems become more sophisticated, they require an ever increasing staff of highly sophisticated personnel" (p. 10). As a result, he saw no savings in actual library operations. Instead, he believes that management is engaging in the game of "musical chairs," in so far as staff allocation was concerned.

Managerial dilemmas extend beyond whether or not technology reduces labor costs. Oberg (1992) found that a growing consensus suggests that the roles of librarians and support staff must be redefined in the face of the rapid changes resulting from automation. The rapidly changing library workplace requires that leadership copes with the tensions created by such aspects as task overlap in which paraprofessionals perform tasks that had been handled by librarians in the past. They were asked to perform duties and assume responsibilities brought on by automation, but without the status or the salaries accorded librarians. As found by Oberg, "the emergence of a paraprofessional category of library employment has been largely uninhibited by associational policy or guidelines" (p. 100). Oberg stipulated that, in the realigned library structure, paraprofessionals were assigned reference and information desk duties, perform a variety of systems work, and catalog most of the books added to the collections. In the brief period since the advent of the Online Computer Library Center (OCLC) they have, in his estimation, come to dominate this work force as well. A redefinition of the roles of librarians and support staff has become a vital issue for library management as they hire and allocate staff to the various task areas. All staff have to be socialized into the values librarians promulgate and defend. Leaders must address differences between support staff and librarians honestly and forthrightly. Pritchard (1995, p. 3–4) reminded managers that a search of common theoretical and methodological guidelines was an elusive but essential beginning.

Kupersmith (1992) listed four components of technology which need to concern library administrators: performance anxiety, information overload, role conflicts, and organizational factors. Establishing a clear mission and set of goals for all employees has been one way of coping with these elements. Hill (1988) perceived that most library tasks remain narrowly focused; that "skills are developed and exercised over relatively small segments of the operation in larger libraries with centralized technical services" (p. 93). However, this was not the case with the administrator. Administration has always been expected to be competent over a range of activities, to be aware of all aspects of librarianship, and to constantly balance the needs of departmental clientele against requirements for the systems as a whole. Bevis and McAbee (1994) investigated managerial tasks from the perspective of involvement with staffing changes. Tasks related to the introduction of automation having an impact on personnel included: reorganization of staff; making changes in job descriptions; making changes in job classifications and salaries; and shifts in staff responsibilities. Making this transition requires an understanding to fill in missing elements that must be acquired whether through new hires for specialized slots, shifting staff into areas consistent with their knowledge and skills, or retaining.

Morris (1991) recommended that the automation of academic libraries ideally would include an administrative commitment to a time and cost study of technical services. This could pinpoint the time spent at various tasks as automation expands and allows the assignment of costs within each of the technical services centers. This kind of research would elucidate the staff time spent at manual activities and then project personnel needs and the need for a reallocation of human resources to deal with the new technology being introduced. Also, in planning special projects, the time data allows for more certain estimates of staffing needs and patterns. Although Morris recognized that such a time/cost study was a difficult assignment, its benefits in offering an insight as to where technical services costs originate and a means of reviewing and improving services, were worthwhile.

Woodsworth et al. (1989) predicted that the academic library of the future will be marked by fluidity and complexity. Teams will form into user-oriented clusters and reform as necessary. Because of this dispersal of staff, administration of the library will be more complicated than before. "In order to maintain their user-driven service orientation, administrators will be increasingly accountable to market needs" (p. 35). They will participate to a greater degree in the development of information policy on campus. "Internally, the role of the administrator in the area of staffing will be to facilitate, coordinate, and orchestrate the work of each component" (p. 35). Further, administrative functioning has varied in its responses to changing environmental conditions. Examples given included knowing when staff skills should be supplemented by "knowbots", expert systems that help define needs, facilitate information access, and tailor information packages to suit the library's clientele and recognizing needs for discipline-based subject expertise, specific to the institution. Decisions have been made as well on when one contracts out for routine operations rather than adding personnel or reassigning present staff.

The Administrators Role in Automating and Staffing the Academic Library
Marchant and England (1989) looked at changing management techniques as libraries automate. They found that, to be effective, library administrators must

manage people and information differently. Many libraries have still managed "using the same model used by industrial age mass production plants. "If we wish libraries to function effectively, they must adapt to a more appropriate mode" (pp. 477–478). In the "information era institution," teams of workers will control information and share in the decision making process, through a participative style of management. Complexity will increase because specialization and diversity will increase. Administrators will be required to make more frequent and rapid decisions, more frequent and rapid innovation, and more rapid, continuous, and wide-ranging information acquisitions. All of this holds true in hiring and allocating staff. Martin (1994) also saw library activities as unique and unamenable to assembly line approach. The increasing reliance on statistics and comparative cost studies such as proposed by Morris tend to ignore this factor according to Martin. The continuing transfer of many activities formerly considered professional in nature (e.g. copy cataloging, acquisitions, interlibrary loan, circulation) to nonlibrarians, and the actual reduction in the total staff members as detailed by Goudy (1993) has pared the personnel share of the library budget from 60 to 50%. This has mandated a rethinking of the way services should be provided accompanied by more of a stress on using staff better, (e.g. getting more work from the existing staff rather than seeking new hires). Retraining provides another means of securing more work from a down-sized staff. Corbin (1991, p. 31) stated that most administrators recognize the need for additional staff training as they introduce new technology.

Montague (1993) explored the role of the administrator in staffing procedures as falling into three major phases: the developmental phase, the operational phase and the integrative phase. Management reactions to these phases in the progression of library automation, as well as typical problems and failures attributed to technology, were characterized in this far-reaching study. During the developmental stage, for example, when the design of work patterns was paramount and few guidelines existed in such areas as role definition between librarians and paraprofessionals, the administrator was expected to play referee and interpreter in these domains. It was stated that the introduction of change, to be effective, requires a management dedication to properly training and motivating the staff so that they fully understand the system, its role in the mission of the library, and the manager's role in making it run.

In a survey conducted by Heaton and Brown (1995, p. 28), staff in a academic library were queried about the role of administration in the automated workplace. The questions related to leadership, goal setting, flexibility, and support. Comments elicited by the respondents suggested lack of communication, lack of encouragement, lack of knowledge, and lack of support. The strong relationship between time and staffing levels was especially apparent. Insufficient staffing or maladaptive allocation of staff resulted in the lack of time available for learning and using the technology.

Finally, Veaner (1993) noted that in some institutions, particularly in their administrative functions, automation was extremely threatening to established departmental structures. It was consistently reported that the political and emotional aspects of such issues as staff realignment processes were more aggravating in system conversion than the technical aspects. The rapidity of change and its consequences for staffing caught some library leaders unprepared, requiring, in effect, reeducation for both management and labor if transitions were to be smooth and efficient.

Staff Hiring, Allocation Policies and Service Provisions

Staffing Considerations for the End-User

Martin (1989) pointed out that libraries are not for librarians but consist of collections and services for users. While not replacing the traditional functions of the library for its users, the new information technologies are "an add-on costing more in time, staff, and equipment, but the value is considerable" (p. 398). Hauptman and Anderson (1994) stated that staff with the appropriate technological expertise for servicing and accessing collections in order to serve users must be trained or hired. The best qualified people must be sought, "especially those who have a vision for new and innovative services using existing resources and technologies supplemented by new technologies and equipment" (249). It was noted too that cost savings in staffing stemming from automation cannot be expected, because, while some time may be saved in the exercise of routine technical tasks, much more effort will have to be devoted to assisting patrons in their use of the new technologies and in maintaining equipment for public use. Hauptman and Anderson also suggested that an important component of personnel planning is listening to what users say about service.

Kliem (1996) stipulated that organizations which shift to a client/server interface must pay attention to this human dimension. This, however, requires greater cooperation between systems staff and end users as well as training for both groups. Pezzulo (1993) held that watching actual usage patterns will give the librarian valuable information. Translating that information into what is needed for students and faculty to relate efficiently with the technology is the real key to effective library usage. Several suggestions were offered as to how patrons might be instructed in the use of library systems. These included having clear instructions available at multiple levels of expertise for each of the resources and otherwise increasing the "comfort" level in the information search process. If we revert back to pre-automation days, the old image of the librarian as the intermediary between information and the learner, whether student or faculty, was discussed. This function is now perceived as a compromise. Intermediation is beneficial for the inexperienced searcher, but the intermediary needs to guide the student toward self-sufficiency. Therefore, when hiring and allocating staff, attention should be directed toward ensuring that personnel are available to educate users on how to access information on their own. Positioning oneself to be the "keeper" of books and technology situating the librarian as gatekeeper, blocks access to information. When considering new hires or retraining existing staff, consideration should be directed toward skills that focus on the role of librarians as teachers.

Special Populations and Remote Sites
Service needs were explored by Martin (1989). They contended that pressure is being placed on academic libraries to evolve in order to serve new structures. Examples were given, such as colleges and universities extending their services to adult learners and who have determined that the establishment of remote sites or campuses is a positive way to reach this population. As a result, the library must identify the appropriate way to deliver information services to these remote sites; new information technologies including telefacsimile, microwave, and satellite links can be utilized to achieve this objective" (p. 376). Often, new organizational structures, effected through hiring practices, staff assignments in relevant positions, or extensive retraining may be the only way to cope successfully with the change.

The Boston Library Consortium (1986) issued a report on several parameters critical for the leader of a library wishing to introduce innovative technologies. The BLC stated that "it is not wise for the library to pull too far ahead of its parent institution's culture and tendencies." Libraries can, if resources are available, install many interesting technologies, databases, and so forth, "but if the users are not ready to accept them, the library will not succeed with those innovations." Another problem with the latest technology and its introduction, is the lack of sufficient knowledge of information technologies on the part of librarians, even those who are recent graduates of library schools. They maintain that courses in new technologies become obsolete almost as rapidly as hardware, thereby hampering efforts to train users to access the latest automated innovations.

In the experience of most researchers, new technologies in the academic library does little to reduce labor costs. On the other hand, it may well be justified to improve services to patrons. Bocher (1993) discerned that better service should be a major goal of any automation effort. For example, circulation and catalog maintenance are labor intensive activities regardless of a library's size. Automating these two operations can free valuable time which can then be invested in user services. But as Horney (1987) asserted, because automation often costs significant amounts for equipment and operations, considerable savings in salaries may be needed to balance expenses. As a result, user services may suffer if low-skilled paraprofessionals take the place of more expensive staff, if fewer workers mean less time to devote to direct user contact, or if increased pressures on a down-sized staff translates into job stress and dissatisfaction and that affects relations with clientele.

Veaner (1993) directed his attention to the dramatic rise in expectations of patrons, especially in the academic community, with the introduction of new technology. In his view, libraries are caught between their perceived failure to utilize to the fullest extent the wonders of automation, while oftentimes lacking the resources for implementing the most desired systems, having in many instances the need to reduce job slots or alter staff assignments and, still confront the "lukewarm commitment of users to undertake the hard work required to learn how to access those systems" (p. 51). Veaner pointed out that much of what was in the library cannot be readily utilized because it was inaccessible because "state-of-the-art staffing does not yet permit far deeper access through librarian-negotiators and patrons at terminals interacting with large and deeply indexed databases" (p. 51). As long as the academic library devotes major portions of its budget and staff to "housekeeping and internal technical processing," it will not realize its full potential as a disseminator of information to its clientele. He also noted that nothing in the educational system forces people to use a given resource; people use the resources that are "effective, responsive, and economical." Staff policies must reflect the realities of present-day library utilization for "patrons growing up in the computer era will not patiently interact with library systems geared to nineteenth-century methods" (p. 51).

Daniels (1995) also stressed the importance of user demands on staffing practices in academic libraries. During implementation of new systems and the consequent extra work represented by the change, existing members of the staff find themselves burdened with heavier work loads, allowing less time to attend to user needs. Even when the systems are in full operation, their very effectiveness and the fact that they allow staff to do so much more, creates more work. Too, users "are becoming more aware of what computers can do for them, and so the level of demand at counters has increased. The system has shown them the full range of information sources available to them and they desire access" (pp. 3–4). While temporary staff may be hired at particularly difficult times, this is not always a viable option. Longer work days and increased workloads are the more usual course.

Whether or not technologies permit more personal contact and communication with users was a point of debate. Daniels reported that some staff enter library work in order to interact with people. They find that automation has, to a certain extent, constrained this interaction as it pertains to users. Employees who were motivated more by "speed and efficiency and providing a fast, accurate service rather than dealing with the clientele are more apt to be satisfied with new technology" (p. 5). However, depersonalization can potentially be a real problem, especially in large academic libraries. Jui (1993) felt that automation would result in libraries and their staffs having more time to examine and deal with other issues, including those revolving around human interfacing. It was stated that many needs of the underserved in the community have been under-addressed by libraries of all types and sizes. Technology provides the opportunity for librarians to reach out, to meet user needs for information which stretch far beyond the usual constituency of libraries. Dyer, Fossey and McKee (1993) found that automation leads to a reduction in contact with library users. An example used was automation which allows students to reserve items without the help of staff. Removing staff from such "public" tasks frequently allows for their reassignment to more "routine" housekeeping tasks such as shelving. It was claimed by Sykes (1986) that "those who find job satisfaction in the rapid no-nonsense provision of research information to users are satisfied with automation; those who enjoy public contact might resent the loss of interaction due to the computerized operations" (p. 56). Technology itself may also preclude many options on the part of administration in the allocation of employees to preferred slots. However, various functions may be resisted or ignored if people are unhappy about their jobs, leading to loss of motivation and, consequently, a decline in satisfactory user service.

Dyer, Fossey and McKee (1993) found that some operators of automated circulation control systems have complained of the "dehumanizing effect of the work. With manual systems, they found that they were more able to communicate with the user than with an automated system where they feel like check-out operators, processing both books and people" (p. 10). It was pointed out that this is not entirely the fault of the automated system. "More information about the borrowers and books can be given on the automated than a manual system" (p. 10). This can help to personalize the service much more if utilized properly, thus working to the advantage of both the employee and the user Robinson and Robinson (1994) went beyond questions relating to job design and satisfaction among staff and patrons. They believed that even the usual investment of time in tracking the uses patrons make of library services, thereby deriving valuable information about library performance from output measures, is no longer adequate. Rather, they held that it is necessary to re-examine what products and services the library has to offer the customer and then identify in detail the resources (inputs) that go into each of these products and services. Increased use of automation has altered these dimensions considerably. The reallocation of existing resources, particularly staffing, in order to introduce new services in response to customer demands, to capitalize on technological advances, to continue to provide products and services which have a proven track record, and to streamline internal library operations, represent a difficult task for the administrator. Becker (1985) asserted that staffing for maximum service delivery is very difficult. For example, the staff must include at least one person technically proficient in each of the products and services offered. "The user population will not seek help from any warm body simply because it works in the library. They expect a support person whose knowledge is more than elementary" (p. 16). The ratio of customers to users is another major concern. Constrained budgets preclude large staffs, but if the staff is spread too thin, it is impossible to provide high quality support to everyone.

Arnold, Collier and Ramsden (1993) reported in depth on the benefits of the automated academic library for students and faculty. "Conventional library support is limiting and inflexible; technology will help users and staff produce higher quality work which is better informed, better research, and more up to date" (p. 3). Further, the information explosion renders conventional libraries "totally unable to manage the volume of information which must be harnessed and made available to higher education" (p. 3). Only electronic storage, transmission, and retrieval, supported by an adequate and well-trained staff, can cope with this challenge. But as Hauptman and Anderson (1994) stated, "patrons do not fully utilize what is available because of inadequate grounding in technology" (p. 249). It was evident that "training and sensitizing of both staff and users will have to be dramatically increased in the future if libraries are to provide the service necessary in an age when no single product is more important than information" (p. 249). Overall, it was held that users are perceived as extremely appreciative of technology and the services it supports. They believe that more technology will lead to even better service. But the quality of the technology accorded to patrons is seen as highly correlated to a well-trained and adequate staff, capable of introducing users to systems and, when necessary continuing to act as a mediator between the user and the system.

RESEARCH DESIGN AND METHODOLOGY

This study used a survey to investigate the effects of automation on hiring practices and staff allocations in four-year and two-year, public and private academic libraries in Tennessee. No distinctions were made between small, medium, and large academic libraries, so the study's population included the total number of four-year and two-year academic library directors located in Tennessee as listed in the 1995–96 *American Library Directory*.

Instrumentation
To determine the effects of automation on hiring practices and staff allocations in the academic library, the Impact of Technology on Library Personnel Survey was developed (Appendix A). The survey consisted of 20 questions with a cover letter detailing instructions for completion of the Survey. The survey was divided into two major parts. The first part asked for the respondents name (which was optional), the name of the library in which he or she worked, and the respondent's position within the library. The second part, included questions about "Staffing and Computerization." Questions in this section could be answered with a "yes", "no", a check mark (/), or a short response. Some questions permitted the respondent to explain his or her answer in more detail. The survey also provided a statement indicating willingness to participate or not participate in the study. Items were coded for utilization in computerized data analysis. To ensure the instrument's validity and reliability, the survey was first mailed to library directors at Belmont University, David Lipscomb, Fisk University, Trevecca Nazarene College, Vanderbilt University, and Tennessee State University. Five of six (84%) of the library directors replied, providing comments and suggestions relating to the clarity of the survey. The data were gathered and analyzed using a t test, percentages, and proportions testing procedures. Revisions were made as a result of the respondents' recommendations and/or suggestions and a second draft of the instrument was constructed. The second draft was then sent to six different academic library directors for further validation of the instrument. These institutions were: Austin Peay State University, Middle Tennessee State University, Lane College, Tennessee Technological University, University of Memphis, and Lambuth University. Five of the six again responded and the survey was finalized. The data were gathered and analyzed using the same statistical procedure.

A third mailing (Appendix B) was then sent to all 56 library directors to be surveyed. Additional mailings and follow-up calls were made, as necessary, in order to maximize return rates.

Table 1. Assessment of the Number of Professional Librarians Before and After Automation.

Departments	*t*	*df*	*p*
Acquisitions	1.36	25	0.19
Cataloging	0.65	25	0.52
Serials	0.33	23	0.75
Circulation	0.44	25	0.66
Reference	1.64	30	0.11

Discussion

Forty-four (78%) of the library directors surveyed completed and returned the survey instrument. The data gathered were placed into appropriate cell ends for analysis and treated with the t test to determine whether there were any statistically significant differences before and after automation.

Hypothesis 1: There will be no significant differences at the 0.05 level in the numbers of professional librarians in academic libraries studied before and after an academic library becomes automated.

The data which responds to this hypothesis was contained in Table 1.

The areas to which the two-tailed treatment were applied emanated from the library's: acquisition, cataloging, serials, circulation, and reference departments. First, the total number of professional librarians was determined both before and after automation. Then, the two-tailed t test treatment was applied to the librarians associated with each of the five departments previously mentioned. No significant difference was found at the 0.05 level in the above comparisons, and the hypothesis was upheld.

Hypothesis 2: There will be no significant differences at the 0.05 level in the numbers of support staff in the study of academic libraries before and after automation.

The data which responds to the preceding hypothesis is contained in Table 2.

Hypothesis 3: There will be no significant differences at the 0.05 level in the number of "other" professional staff (computer programmers, computer specialists, etc.) in academic libraries before and after automation.

The researcher was not able to execute the two-tail *t* statistical treatment on the data necessary to accept or reject Hypothesis 3. Respondents provided

Table 2. Assessment of the Number of Support Staff Before and After Automation.

Departments	*t*	*df*	*p*
Acquisitions	0.94	26	0.36
Cataloging	0.50	24	0.62
Serials	0.83	22	0.42
Circulation	1.50	25	0.15
Reference	0.27	21	0.79

No significant difference at the 0.05 level was found in the above comparisons, and again the hypothesis was upheld.

insufficient data for analysis and evaluation. However, gleaned information is presented in Table 3. This table shows that 19 respondents (43.2%) provided no response. Three (6.8%) respondents had hired one post automation computer specialist, while two (4.6%) individuals had hire one new staff member. Finally, 20 (45.5%) respondents indicated that no additional staff in this category had been hired.

Hypothesis 4: Hiring policies, for example, specific skills will not change subsequent to academic library automation.

The study included questionnaires about hiring practices to include items (11): Are computer skills presently required of your support staff? (14): Have job descriptions changed to specify the computer skills needed for professional librarians? and (15): Have job descriptions changed to specify the computer skills needed for support staff? All respondents indicated that while some employees had computer skills cited above before automation, they were not required and/or prerequisite to employment and/or keeping the job until automation occurred.

Table 3. Other Professionals Added.

Staff Hired	Frequency	Percentage	Valid Percent
0	20	45.5	80.0
0.5	1	2.3	4.0
1	3	6.8	12.0
2	1	2.3	4.0
0	19	43.2	Missing
Totals	44	100.0	100.0

Table 4.   Computer Skills Required of Support Staff.

Response	Frequency	Percent	Valid Percent
Yes	39	88.6	92.9
No	3	6.8	7.1
No Response	2	4.5	Missing

The nature of the study responses dictated that the two-tailed t-test of statistical significance should not be employed. Available related data are presented in Tables 4, 5, and 6.

Table 4 provides a breakdown of the question to respondents to whether computer skills were presently required of support staff.

Thirty-nine (92.9%) respondents affirmed that computer skills were required of support staff, while three (7.1%) library directors required no previous knowledge of the computer. Two (4.5%) respondents did not provide an answer to the question.

Table 5 provides the results of the participants responses to whether job descriptions have changed to specify the computer skills needed for professional librarians.

Thirty (71.4%) of the study participants now require job skills commensurate with job descriptions for the professional librarian. Twelve (28.6%) respondents reported that they did not require computer skills for the professional librarian including requisite computer skills.

Table 6 shows the results of whether job descriptions have changed to specify the computer skills required of support staff.

Thirty-four (81.0%) of library directors responded that job descriptions have changed to specify needed support staff employee computer skills, while only eight (19.0%) responded that no changes had occurred in the way they hired staff. Two respondents did not provide a response to this item.

Table 5.   Job Descriptions Changed to Specify Computer Skills Needed By Professionals.

Response	Frequency	Percent	Valid Percent
Yes	30	68.2	71.4
No	12	27.3	28.6
No Response	2	4.5	Missing
Totals		100.0	100.0

Table 6. Job Descriptions Changed to Specify Computer Skills Needed By Support Staff.

Response	Frequency	Percent	Valid Percent
Yes	34	77.3	81.0
No	8	18.2	19.0
No Responses	2	4.5	Missing

Summary Analysis of Research Questions

Research Question 1: Is there a significant difference in the numbers of professional librarians employed before and after an academic library an academic library becomes automated?

Survey questionnaire items 13 and 19 were developed to respond specifically to this research question: How many professional and support staff have you added to your library since automation. Table 7 provides data which responds to this research question.

Table 7 reflects the data analysis results for item 13. This material shows that 7 (15.9%) of the study respondents reported a post automation increase in professional staff. Companion questionnaire item 19 which indicates the number of professional and support staff in each library department, before and after automation is carried in the data provided in Table 7 and Table 8.

Research Question 2: Is there a significant difference in the numbers of support staff employed before and after an academic library becomes automated?

The data contained in Table 8 relate to the statistical breakdown of support staff provided by study respondents.

Result of analysis indicates that 20 (60.6%) respondents hired no new support staff subsequent to automation. Ten (29.4%) participants employed one new

Table 7. Professional Staff Increase Post Automation.

Response	Frequency	Percent
Increase	7	15.9
No Change	37	84.1
Totals	44	100.0

Table 8.   Support Staff Added After Automation.

New Hires	Frequency	Percent	Valid Percent
0	20	45.5	60.6
1	10	22.7	29.4
2	3	6.8	8.8
4	1	2.3	2.9
No Response	10	22.7	Missing
Totals	44	100.0	100.0

support staff, three (8.8%) hired two additional employees, while one (2.9%) respondent brought four new individuals on board. Ten (22.7%) library directors did not provide a response to this questionnaire item.

Research Question 3: Is there a significant difference in the numbers of "other" professional staff employed before and after an academic library becomes automated?

Twenty (80.0%) of the respondents added no "other" professionals (computer specialists, computer programmers, etc.) to their library staff, while one director (4.0%) added a half-time employee. Three (12.0%) added one "other" staff member, with one (4.0%) respondent adding two employees in this category. Nineteen library directors (43.2%) did not respond to this specific questionnaire item.

Research Question 4: Do job descriptions, content, and status alter when an academic library becomes automated?

Table 9 and 10 reflect the data collected from questionnaire items 14 and 15 concerning the manner in which job descriptions had changed for professionals and support staff, respectively.

Table 9 reports that 30 (71.4%) of the study respondents affirmed that job descriptions had changed for professionals staff, while 12 (28.6%) responded that no change had taken place with respect to altering professional job descriptions. Two (4.5%) did not respond to the questionnaire item.

Table 11 reports the data analysis of the study's respondents to questionnaire item 15 regarding altering job descriptions for support staff.

Table 11 indicates that 8 (19.0%) of the study respondents reported that job descriptions altered for support staff, while two (4.5%) indicated no response.

Table 9. "Other" Professional Hired Post Automation.

New Hires	Frequency	Percent	Valid Percent
0	20	45.5	80.0
0.5	1	2.3	4.0
1	3	6.8	12.0
2	1	2.3	4.0
No Response	19	43.2	Missing
Totals	44	100.0	100.0

Questionnaire item 10 indicated whether staff reductions occurred due to automation. Figure 1 indicates that two (4.5%) respondents had reduced library personnel, while 10 (22.7%) had increased the number of library employees. Forty-two (95.5%) of respondents did not respond to the questionnaire item.

Questionnaire item 10 also indicated whether staff reductions occurred due to budget cuts. Figure 2 provides the results of the participant's responses to whether budgetary factors influenced an increase or decrease in staffing patterns. Five (11.4%) of study respondents reported a reduction of library personnel due to budgetary cuts, while thirty-nine (88.6%) did not respond to the query. At the same time 6 (13.6%) of the respondents reported an increase of staff linked to budget cuts, while (86.4%) did not respond to the questionnaire item.

Questionnaire item 17 indicated what would be the most immediate impact on future budget-cutting on library staffing patterns. Figure 3 indicates that seven (19.4%) participants reported that fewer professionals would be hired, two (5.6%) indicated that fewer support staff would be hired, two (5.6%) responded that they could afford more automation, five (13.9%) provided numbers that indicated less automation was possible, and 15 (41.7%) offered that a cut back in services would be more likely. Eight (18.2%) of the participants did not respond to the questionnaire item.

Table 10. Job Descriptions Altered for Professionals.

Response	Frequency	Percent	Valid Percent
Yes	30	68.2	71.4
No	12	27.3	28.6
No Response	2	4.5	Missing
Totals	44	100.0	100.0

Table 11. Post Automation Job Descriptions Altered for Support Staff.

Response	Frequency	Percent	Valid Percent
Yes	34	77.3	81.0
No	8	18.2	19.0
No Response	2	4.5	Missing
Totals	44	100.0	100.0

FINDINGS AND CONCLUSIONS

This study was undertaken with the expressed objective of determining the impact of automation on hiring activities and staff apportionment in two-year and four-year academic libraries in the state of Tennessee. The intent was to examine in depth ways in which automation has altered organizational structure as experienced by the forty-four library directors taking part in the study. Adaptation to the rapid computerization of academic libraries inevitably involves re-configurations among employees, both credentialed professionals and support personnel. Automation forces the recognition that established structural and procedural practices, long taken for granted, are no longer viable. Traditional lines of responsibility and authority have blurred, even

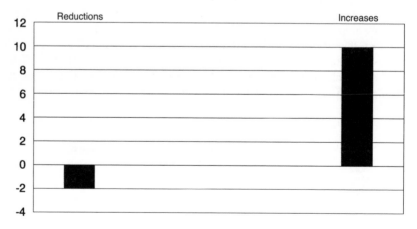

Fig. 1. Libraries Reporting Staff Reductions or Increases Due Mainly to Automation.

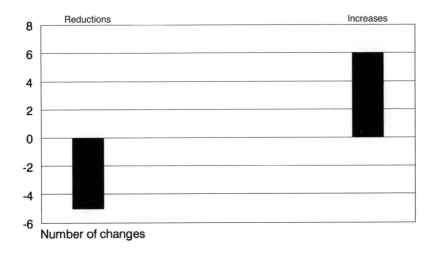

Fig. 2. Libraries Reporting Staff Reductions or Increases Linked to Budget Cuts, Following Automation.

as technological forces dictate changes in how the academic library perceives its mission to its clientele.

The consensus of the literature tended to indicate that automation may have a positive or negative effect on staffing procedures in the academic library. Most strategies utilized in the process of computerization assume the efficacy of a role/task approach to staffing. More personnel positions may be added in particular areas to facilitate the demands of the new technology, and drop those that are no longer deemed essential could be dropped. Similarly, more tasks could be added to a job description, tasks once handled by separate entities combined, or tasks could be dropped as no longer necessary to library operations. Present day budgetary constraints on academic libraries, the pressure for increased productivity and accountability, in part wrought by automation's presumed ability to better utilize available resources in a rational way, and the human factors inherent in technological changes emphasize the importance of individual roles and responsibilities. However, the task perspective, stressing functions and processes, also necessitates clarification in an era of automation. The analysis and alignment of role and task, personnel and responsibilities are mainly the province of the library director. Through specific strategies in the hiring and allocation of staff members, the academic library director coordinates, organizes, and manages the many components related to the human element of automation.

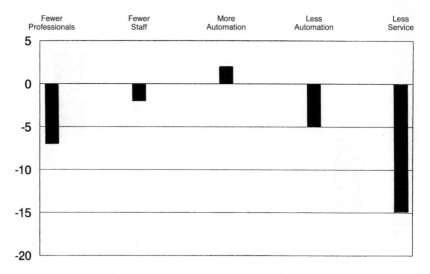

Fig. 3.   Impact of Budget Cuts.

Findings

There were four hypotheses articulated and presented and tested during the course of this study. The hypotheses were aligned to the research questions, the findings for which were examined above. In brief, Hypothesis 1, in its null form, held there would be no significant differences in the numbers of professional librarians in academic libraries studied before and after automation. The null was accepted at the .05 level of significance. Hypothesis 2 stated in its null form that there will be no significant differences in the numbers of support staff in the study of academic libraries before and after automation. This hypothesis was accepted at the .05 level of significance. Hypothesis 3, in its null form stated that there would be no significant differences in the numbers of "other" staff employed in academic libraries before and after automation. While the previous two hypotheses permitted the utilization of the two-tailed statistical treatment on the data necessary to accept or reject Hypothesis 1 and 2, respondents to Hypothesis 3 failed to offer the necessary data for analysis and evaluation through this methodology. Nineteen directors (43.2%) did not respond. Three respondents (6.8%) had hired one post automation specialist, while two (4.6%) of the directors had hired one new staff member. Considering only those directors who responded to this question, it is deemed that significant numbers of "other" professional staff members, those directly employed in the computer-related tasks, had not been hired in significant numbers.

Hypothesis 4 revolved around hiring policies following automation in the academic library. The questionnaire contained three items (11) Are computer skills presently required for your support staff?, (14) Have job descriptions changed to specify the computer skills needed for professional librarians?, and (15) Have job descriptions changed to specify skills needed of support staff. All the respondents indicated that computer skills were necessary after automation. However, some employees were already proficient in computer usage before they became a necessary factor in employment after the introduction of automation in their libraries. Further, before automation, computer skills may not have been a requirement, were not a condition for employment, and/or not judged a requisite for job retention. The data generated by these questions were displayed in three Tables: 4, 5, and 6. Table 4 provided a breakdown of whether or not computer skills were a requirement for support staff following automation. Thirty-nine of the directors (88.6%) indicted that computer skills were indeed a necessity for support staff; three directors (6.8%) stated their academic libraries had no such requirement; two respondents (4.5%) did not answer the questionnaire item. The comparative weight of the affirmation of the need of computer skills for support staff in the academic library supports the thesis that such skills have become a condition for employment in the automated environment.

Table 5 supplied the results on whether job descriptions have changed for professional librarians to better reflect the computer skills mandated for these employees. Thirty directors (68.2%) indicated that computer skills commensurate with the job descriptions for professional hires were required; twelve (27.3%) of the respondents reported they did not require computer skills for professional librarians corresponding to job descriptions. Two participants failed to respond (4.5%). The data suggests that computer knowledge is held to be of importance for professional librarians but not to the extent it has become a job requirement in the hiring of support staff.

Table 6 provided the specifics on changes in job descriptions for the support staff in line with the computer skills required of them. The majority of directors, 34 respondents (77.3%) stated that job descriptions had altered to specify what computer skills were required of support staff; 8 directors (18.2%) held that no changes had occurred in the way they hired support staff. Two participants (4.5%) did not respond to the question. Once again, evidence suggested that job descriptions, in particular those for support staff, had changed dramatically following introduction of computers. It is reasonable to assume that the prerequisite of computer skills among support staff has significantly impacted on recruitment and hiring practices as well as retraining programs to provide for new priorities and expectations.

Conclusions

For much of the history of automation in the academic library, concern has been to chart the technological aspects of rapid computerization. Only recently has interest been manifested in the human dimensions of automation in this setting. Information revolving around personnel issues generated by automation is, as yet, quite limited. Therefore, the implications for studies of changes in hiring practices and staff allocations in academic libraries are of value to library directors and others responsible for implementing new configurations and utilization of staff in ways demanded by new technology and cost effectiveness. Increased productivity and customer satisfaction are the two goals most notably connected to new technological advances. To accomplish these ends, shifts in job classifications, duties, and responsibilities are oftentimes required.

As more operations and services in the academic library are computerized, the design of individual jobs and the interrelationships between jobs, between professional and support staff, and between internal and external pressures for economy and efficiency are impacted. The current study sought to detail how, in certain specific areas, automation has, or perhaps has not, changed he face of the academic library. For different reasons, both professional librarians and support personnel have feared their positions would be "endangered" or downgraded because of the new technology. The findings indicated that neither group has gained or lost much ground due to automation. Nor is the expectation that automation signals the rise of the "specialist" as revealed throughout this research. While hiring and retention in a position is more dependent on the possession of computer skills than in the past, the skills have not translated into the large-scale demand for a specialist to fulfill the needed functions.

Certainly, job descriptions, in particular those related to support staff, have undergone changes due to incorporating the necessary knowledge. But this has not resulted in major shifts in the numbers or types of staff employed. Any direct relationship between automation and re-configuration of personnel is difficult to discern at the present time. Only in the expressed need for staff, especially support staff, to possess computer skills has the requirements for personnel been influenced to any real degree. Expertise of a radically different kind, slanted toward conceivable new roles and knowledge is not perceived as a felt need by the library directors currently.

It appears that automation has not drastically changed personnel practices in the academic library and has not significantly altered ratios between professional and support staff. But it has, to a certain extent, led to rewriting of individual job descriptions which take into account the needs of the new technology. Moreover, little was discovered in this study about staff working across job classifications or whether or not traditional hierarchical lines of responsibility and task allocation have been breached by automation, nor is there evidence of staff reductions in various areas, different ways of configuring, the organization or demands for flexibility in staff allocations. However, it is clear that we have only begun to understand the impact of automation on the academic library staff and organization.

SELECTED REFERENCES

Arnold, K., Collier, M., & Ramsden, A. (1993). ELINOR: The electronic library project at DeMontfort University Milton Keynes. *Aslib Proceedings, 45*, 3–6.

Atkinson, H., & Stenstrom, P. (1984). Automation in austerity. In: J. Harvey & P. Spyers-Duran (Eds), *Austerity Management in Academic Libraries* (pp. 281–282). Metuchen, N.J: Scarecrow.

Auld, L. W. S. (1987). The King report: New directions in library and information science education: A close look at a controversial study. *College and Research Library News, 48*, 178.

Baker, N. (1996, October 14). Letter from San Francisco: The author vs. the library. *New Yorker*, 51–52.

Balbach, E. D. (1989). Personnel. In: B. P. Lynch (Ed.), *The Academic Library in Transition: Planning for the 1990s* (p. 311). New York, New York: Neal-Schuman.

Bartlett, V. (1995). Technostress and librarians. *Library Administration and Management, 9*, 226–229.

Becker, J. B. (1985, October/November). How information centers fail. *ASIS Bulletin*, 16–17.

Berman, S. (1987). Automated cataloging: More or less staff needed? *Library Journal, 103*, 415.

Bevis, M. D., & McAbee, S. (1994). NOTIS as an impetus for change in technical services departmental staffing. *Technical Services Quarterly, 12*, 29–43.

Birdsall, D. C. (1991). Recruiting academic librarians: How to find and hire the best candidates. *The Journal of Academic Librarianship, 17*, 276–283.

Bocher, B. (1993, February). Small automated library systems. *Computers in Libraries, 13*, 26.

Bonta, B. D. (1990). Library staffing and arrangements for CD-ROM service. *Inspel, 24*, 5–13.

Bosseau, D. (1992). Confronting the influence of technology. *The Journal of Academic Librarianship, 18*, 302–303.

Boss, R. W. (1984). *The Library Manager's Guide to Automation*. White Plains, Knowledge Industry Publications, 7–8, 104.

Boston Library Consortium (1986). Managing technological change. (interim report): Boston, MA: n. p.

De Bruijn, E. (1986, December). The effect of automation on job duties, classifications, staff patterns, and labor costs. Vancouver, BC: University Press.

Cain, M. E. (1986). Research libraries in transition: Managing in the university setting. In: M. M. Cummings (Ed.), *The Economics of Research Libraries* (pp. 182–186). Washington, DC: Council on Library Resources.

Cline, H. F., & Sinnott, L. T. (1983). *The Electronic Library: Impact of Automation on Academic Libraries*. Lexington, KY: Heath.

Convoy, B. (1982). *The Human Element: Staff Development in the Electronic Library*. Philadelphia, PA: Drexel University.

Corbin, J. (1991, November). Automation and non-professional staff – the neglected majority. In: S. Sykes (Ed.), *Serials* (p. 39).

Craghill, D., Neale, C., & Wilson, T. D. (1989). *The Impact of IT on Staff Deployment in U.K. Public Libraries*. London: British Library Board.

Dakashinamurti, G. (1985). Automation's effect on library personnel. *Canadian Librarian Journal, 42*, 343–351.

Daniels, R. J. (1995, January). Effects on non-professional staff of the implementation of computer-based library systems in college libraries: Some case studies. *Program*, 1–13.

De Gennaro, R. (1994, October 15). Library automation: Changing patterns and new directions. *Library Journal*, S8-S16.

De Klerk, A., & Euster, J. (1989, Spring). Technology and organizational metamorphoses. *Library Trends, 37*, 457–468.

Douherty, R. M. (1994). Editorial: On becoming all we can be: A last word. *Journal of Academic Librarianship, 19*, 355.

Dyer, H., Fossey, D., & McKee, K. (1993, January). The impact of automated library systems on job design and staffing structures. *Program, 27*, 1–16.

Goudy, F. W. (1993). Academic libraries and the six percent solution: A twenty-year financial overview. *Journal of Academic Librarianship, 19*, 212–215.

Hauptman, R., & Anderson, C. (1994, December). The people speak: The dispersion and impact of technology in American libraries. *Information Technology and Libraries, 13*, 249.

Haywood, T. (1991). Changing faculty and academic environments within which U.K., U.S., and Canadian library/information schools are operating and the influence this might have on the broadening or narrowing of curriculum development. London: British Libraries.

Heaton, S., & Brown, J. (1995, February). Staff perceptions of incentives and hurdles to the use of technology. *Computers in Libraries, 15*, 28.

Hegarty, K. (1985, October 1). Myths of automation. *Library Journal*, 45.

Heim, K. M., & Moen, W. E. (1990). A survey of information services recruitment: The challenge of opportunity. *Research Quarterly*, 562–566.

Hill, J. S. (1988). Staffing technical services in 1995. *Journal of Library Administration, 9*, 87–103.

Horney, K. L. (1987, January/March). Fifteen years of automation: Evolution of technical services staffing. *Library Resources and Technical Services, 31*, 69–76.

Jui, D. (1993, December). Technology's impact on library operations. *EDRS*, 1–11.

Johnson, P. (1991). *Automation and Organizational Change in Libraries*. Boston, MA: Hall.

Kiesler, S., Obrosky, S., & Pratto, F. (1987). Automating a university library: Some effects on work and workers. In: S. Kiesler & L. Sproull (Eds), *Computing and Change on Campus* (pp. 131–149). Cambridge, MA: Cambridge University Press.

Kliem, R. L. (1996, January/February). Managing the people side of client/server architecture. *Journal of Systems Management, 47*, 24–25.

Koenig, M. E. D. (1993, Fall). Educational requirements for a library-oriented career in information management. *Library Trends, 42*, 277–289.

Kraske, G. (1978). The impact of automation on staff and organization of a medium-sized academic library: a case study. Terre Haute, IN: Indiana State University.

Kupersmith, J. (1992, Summer). Technostress and the reference librarian. *Reference Services Review, 20*, 9.

Langdon, P. (1995, June). Faces of a downsized profession. *Progressive Architecture, 76*, 75–77.

Love, J. P. (1994 January/February). A window on the poltics of the Government Printing Office Electronic Information Access Enhancement Act of 1993. *Journal of Government Information, 21*, 3–13.

Malinconico, S. M. (1994, October 15). People and machines: Changing relationships. *Library Journal*, S2-S4.

Manson, P. (1987, March/April). Automation: Fools rush in – problems and solutions. *Serials, 10*, 108–120.

Marchant, M. P., & England, M. M. (1989, Spring). Changing management techniques as librarians automate. *Library Trends, 37*, 469–483.

Martin, M. S. (1989, July). Information technology and libraries: Toward the year 2000. *College and Research Libraries*, 397–405.

McCrank, L. J. (1986). The impact of automation: integrating archival and bibliographical systems. *Journal of Library Administration, 7*, 61–98.

McDaniel, M. A., & Schmidt, F. L. (1989, Spring). Computer-assisted staffing systems: The use of computers in implementing meta-analysis and utility research in personnel selection. *Public Personnel Management, 18*, 75–86.

McLean, N. (1981). Computerization and library organization. The management of technical innovation in libraries. Proceedings of a conference. Loughborough: Center for Library and Information Management, 1981, 7.

Martin, S. K. (1989, Winter). Library management and emerging technology: The immovable force and the irresistible object. *Library Trends, 37*, 374–382.

Mitchell, M., & Saunders, L. (1993, November/December). The national information infrastructure: Implications for libraries. *Computers in Libraries, 13*, 53–56.

Montague, E. (1993, March). Automation and the library administrator. *Information Technology and Libraries, 12*, 77–79.

Morris, D. E. (1991). Staff time and costs for cataloging. *Computers in Libraries, 36*, 79–85.

Oberg, L. R. (1992). The role, status, and working conditions of paraprofessionals: A national survey of academic libraries. *College and Research Libraries, 53*, 215–238.

Oberg, L. R. (1995). Library support staff in an age of change: Utilization, role definition and status. *ERIC Digest*, 1–5.

Olsgaard, J. N. (1989, Spring). The physiological and managerial impact of automation on library. *Library Trends, 37*, 4.

Palmini, C. C. (1994, March). The impact of computerization on library support staff: A study of support staff in academic libraries in Wisconsin. *College and Research Libraries*, 119–127.

Pezzulo, J. (1993, May). The human interface with technology. *Computers in Libraries, 13*, 20–24.

Prince, B., & Burton, P. F. (1988). Changing dimensions in academic library structures: The impact of information technology. *British Journal of Academic Librarianship, 3*, 67–81.

Pritchard, S. M. (1995, Spring). The ultimate moveable type. *Moveable Type, 2*, 3–4.

Pungitore, V. L. (1986). *Development and Evaluation of a Measure of Library Automation*. Bloomington, IN: School of Library Science, Indiana University.

Robinson, B. M., & Robinson, S. (1994, Winter). Strategic planning and program budgeting for libraries. *Library Trends, 42*, 420–447.

Robison, C. B. (1991). Effects of automation on academic libraries. (Unpublished doctoral dissertation, Oklahoma State University, 1991).

Saunders, L. (1996, May). Changing technology transforms library roles. *Computers in Libraries, 16*, 49.

Schnelling, H. (1992, November). Management issues in library automation: Experiences from German university libraries. Paper presented at the EEC Comett Course, University Libraries and Information Transfer: Automation as a New Organizational Language, Venice, November 23–26th, 1992, 323–341.

Smalley, T. N. (1994). Computer systems in libraries: Have we considered the tradeoffs? *Journal of Academic Libraries, 12,* 356–361.

Smith, K. (1993, June). Toward the new millennium: The human side of library automation. *Information Technology and Libraries, 12,* 209.

Stuart-Stubbs, B. (1994, November). Trial by computer. *Library Journal,* S12–S16.

Sykes, P. (1986). Automation an non-professional staff at the Polytechnic of the South Bank. *Training and Education, 3,* 50–56.

Thompson, R. K. H. (1992, May). Funding for library automation. Paper presented at the National Conference of the Library and Information Technology Association (3rd, Denver, CO, September 13–17, 1–6.

Veaner, A. B. (1994). Paradigm lost, paradigm regained? A persistent personnel issue in academic librarianship. *College and Research Libraries, 55,* 389–402.

Veaner, A. B. (1993). Institutional political and fiscal factors in the development of library automation, 1967–71. *Information Technology and Libraries, 12,* 51.

Wall, C. E. (1985). In quest of automation. *Library Hi-Tech, 3,* 5, 72.

Woodsworth, A., Allen, N., Hoadley, I., Lester, J., Molholt, P., Nitecki, D., & Wetherbee, L. (1989). The model research library: Planning for the future. *Journal of Academic Librarianship, 15,* 132–138.

APPENDIX A

IMPACT OF TECHNOLOGY ON LIBRARY PERSONNEL SURVEY

Respondent's Identification

Name_____(Optional)

Name of Library_____

Position_____

Staffing and Computerization

1. Does your library have the following? (Check all that apply)

    Online catalog   1. _____ yes   2. _____ no

    Local area network   1. _____ yes   2._____no

    Wide area network   1. _____ yes   2. _____ no

    Internet access   1. _____ yes   2. _____ no

    Other (Specify):   _____

2. Identify which of the following functions are automated?

    Circulation   1. _____ yes   2. _____ no

    Cataloging   1. _____ yes   2. _____ no

    Serials   1. _____ yes   2. _____ no

    Acquisitions   1. _____ yes   2. _____ no

3. Is more time spent at the reference desk since automation?

    1. _____ yes   2. _____ no

4. Has the level of online searching changed since library automation?

    1. _____ yes   2. _____ no

    3. _____ about the same   4. _____ don't know

5. How many databases do your users have access to?

    1. _____

6. Have library orientations increased or decreased since automation?

    1. _____ increased   2. _____ decreased

7. Has retraining of professional librarians been necessary following automation?

    1. _____ yes   2. _____ no

8   Has retraining of support staff been necessary following automation?

    1. _____ yes   2. _____ no

9   How many professional library positions were reclassified following automation?

    1. _____ number of positions reclassified

    Indicate new position titles, if any:

    2. _____   3. _____   4. _____

10 If staff changes occurred, was this reduction or increase due to automation, budget allocations, or other factors?

1. _____ reduction due to automation

2. _____ increased due to automation

3. _____ reduction due to budget cuts

4. _____ increased due to budget allocations

5. _____ other (Specify): _____

11. Are computer skills presently required of support staff?

1. _____ yes   2. _____ no

12. How many support staff positions have been reclassified following automation?

1. _____

Indicate below the positions, if any, by name:

2. _____   3. _____

4. _____   5. _____

13. How many professional and support staff have you added to your library since automation?

1. _____ number of professional librarians
2. _____ number of support staff
3. _____ number of technical personnel (other than librarians or support staff)

14. Have job descriptions changed to specify the computer skills needed for professional librarians?

1. _____ yes   2. _____ no

15. Have job descriptions changed to specify the computer skills needed for support staff?

1. _____ yes   2. _____ __no

16. Has the ratio of professional to support staff changed?

1. _____ yes   2. _____ no

If yes, has it:

3. _____ increased number of professional staff

4. _____ increased number of support staff

5. _____ decreased number of professional staff

6. _____ decreased number of support staff

17. What would be the most critical impact of future budget-cutting on library staffing practices?

1. _____ fewer professional librarians

2. _____ fewer support staff

3. _____ more automation

4. _____ less automation

5. _____ cut back on services

6. _____ other (Specify): _____

18. Do you believe that automation has increased or decreased the overall effectiveness of your library?

1. _____ increased   2. _____ decreased

19. Please list below how many professional and support staff positions existed in each of the following departments BEFORE and AFTER automation?

Acquisitions Department		*Circulation Department*	
Before	*After*	*Before*	*After*

1. _____ Professional staff _____   1. _____ Professional staff _____

2. _____ Support staff _____   2. _____ Support staff _____

Cataloging Department		*Reference Department*	
Before	*After*	*Before*	*After*

1. _____ Professional staff _____   1. _____ Professional staff _____

2. _____ Support staff _____   2. _____ Support staff _____

Serials Department	
Before	*After*

1. _____ Professional staff _____

2. _____ Support staff _____

1. Have you had a need to hire "other" professional staff, e.g. computer specialists, programmers, etc.?

   1. _____ yes

   2. _____ no

QUALITY ASSURANCE IN LIBRARY SUPPORT OF DISTANCE LEARNING: INTERNATIONAL PERSPECTIVES FOR LIBRARY ADMINISTRATORS

Alexander L. Slade

INTRODUCTION

As reflected by the literature, librarians have long considered access to library resources to be essential to the quality of distance education programs. Writings on this topic date back to 1965 when Arthur T. Hamlin eloquently argued the case for adequate library support in extension programs at Hearings held by the Subcommittee on Education of the United States Senate's Committee on Labor and Public Welfare (Hamlin, 1965). Subsequently, authors from various countries have emphasized the importance of library services for the quality of distance learning. However, with a few notable exceptions, these authors have been librarians rather than distance educators or academics. The distance education literature has generally been silent on the topic of library services. Borje Holmberg does make reference to the role of libraries in his book *Theory and Practice of Distance Education* (Holmberg, 1989), and authors writing in *Why the Information Highway? Lessons from Open and Distance Learning* acknowledge the contribution that libraries can make to distance learning (Roberts & Keough, 1995). Otherwise it has usually been left to librarians to delineate the need for their services.

Advances in Library Administration and Organization, Volume 18, pages 225–244.
Copyright © 2001 by Elsevier Science Ltd.
All rights of reproduction in any form reserved.
ISBN: 0-7623-0718-8

American library initiatives related to quality in distance learning have been well-documented, especially in relation to the ACRL *Guidelines for Distance Learning Library Services* (Association of College and Research Libraries, 2000). However, initiatives in other countries have received less attention and information about international issues in this area is relatively limited. The focus of this paper is to highlight the role of library services in quality assurance for distance learning outside of the United States, and to examine current and future trends in the provision of library services for distance learning. This discussion is based on the contents of the three annotated bibliographies on library services for off-campus and distance education (Latham, Slade & Budnick, 1991; Slade & Kascus, 1996, 2000).

FACTORS INFLUENCING LIBRARY SERVICES

There are four significant factors that differentiate the United States from other countries in terms of library support for distance learning. These factors pertain to regional accrediting agencies, national guidelines, access to other libraries, and the extended campus model.

The United States is unique in that it is the only country that has regional accreditation agencies to evaluate the quality of post-secondary education. Most American colleges and universities have to deal with these accreditation bodies with regard to library services for distance education. Accreditation reviews are often cited as the impetus for academic libraries in the U.S.A. to enhance services for distance learners. Gilmer (1995) writes that, in recent years, most accreditation agencies have revised their standards with greater emphasis on off-campus library services. Several authors describe the necessity for their institution to consider accreditation standards when planning library services. Caballero and Ingle (1998), for example, describe how the University of Texas at El Paso included accreditation and library standards considerations from the Southern Association for Colleges and Schools in the development of library service in support of distance education.

In other countries, academic institutions tend to be regulated at either the federal or provincial/state level and are not subject to the scrutiny of accreditation bodies. Exceptions are generally found in the case of professional schools that require accreditation from the corresponding professional organization. Outside of the United States, accreditation is not a significant factor affecting the provision of library services for distance learning.

In many ways, the ACRL *Guidelines for Distance Learning Library Services* are designed to complement the accreditation process in the United States. American librarians have been concerned about guidelines for extended campus and distance learning library services since 1967 when the first set of guidelines appeared. The U.S. guidelines were revised in 1982, 1990, 1998 and again in 2000. The ACRL Distance Learning Section, which authored the current version, is actively promoting the guidelines to libraries, educational bodies, and accrediting agencies. Many American libraries use these guidelines as the basis for planning and initiating services for distance learners.

Guidelines for distance learning library services have not received as much attention in other parts of the world. Canada developed a set of guidelines in 1993 and has recently revised them (Canadian Library Association, 2000). However, relatively little has been written about the Canadian guidelines. Australia has a set of guidelines for library services to external students but its version was published in 1982 and, to date, has not been revised (Crocker, 1982). Unbeknown to many, guidelines also exist in the United Kingdom but are seldom cited (Fisher, 1988).

Another factor that distinguishes the U.S.A. from other countries in terms of distance learning library services is access to other libraries. Due to the population density in many parts of the United States, off-campus students are often located within commuting distance of one or more academic libraries. Use of libraries not affiliated with the "home" or "originating" institution has been one of the most common means for distance learners to obtain access to library resources. It has also been a much debated topic in the American library literature, leading to usage of the term "victim libraries" (Dugan, 1997).

In countries with large rural areas, including Canada and Australia, distance learners may be located far from other post-secondary institutions and frequently lack the option of visiting another academic library in person. As a result, library services for distance learners in these countries tend to be provided from the "home" library through postal and document delivery services. In the United Kingdom where population density is great, access to other university libraries is limited due to restrictive policies. One of the findings from a study on the library experiences of postgraduate distance learning students in the United Kingdom was that use of other libraries is a clandestine activity for some students due to time limitations for site visits and the restrictive policies of many university libraries (Stephens, 1998). Public libraries in the United Kingdom are more oriented to serving the adult learner than public libraries in the U.S.A. and, for that reason, distance students in the U.K. often turn to their local public library for assistance when they need library resources for their course work.

A fourth factor that demarcates the United States from other countries in the area of distance learning library services has been the prevalence of the extended campus model of education. This model which involves the use of satellite campuses or remote teaching centers for course delivery has been the primary means of off-campus education in the U.S.A. until recently. While distributed learning via the Internet and World Wide Web is quickly gaining prominence, many off-campus teaching sites are still in use. This model has impacted the provision of library services, focusing more on on-site collections and resources than on remote access to library materials.

Remote teaching centers are not nearly as common in other countries. Instead one finds regional and study centers, especially in countries with open universities. These centers are intended to provide some local support for distance learners in terms of access to audio-visual and computer equipment, study facilities, and tutorial and counselling services. Little or no teaching takes place at these centers and access to library resources tends to be limited. In developing countries, these centers tend to be problematic because they are often the student's only means to obtain library materials for distance learning courses. Few institutions in developing countries can afford to deliver library materials directly to students due to transportation and other difficulties. Public libraries in these countries rarely have materials appropriate to distance education coursework. Common problems with study centers include insufficient funding, poorly trained staff, limited opening hours, inadequate collections that sometimes do not circulate, lack of equipment, and minimal reference materials. Students living in remote areas may be unable to visit these centers on a regular basis. The problems associated with regional and study centers have been particularly well documented in the literature from India (e.g. Gupta, 1997).

THE CANADIAN GUIDELINES

The existence of national guidelines on library services for distance learning is one of the major benchmarks for quality assurance in this area. In this regard, the United States is ahead of other countries due to its long-standing commitment to library guidelines. Canada is the only other country to have a current set of guidelines on distance learning library services.

The *Guidelines for Library Support of Distance and Distributed Learning in Canada* were developed by a committee of the Services for Distance Learning Interest Group of the Canadian Library Association. The Guidelines were published in 1993 (Canadian Library Association, 1993) and revised in 1999 (Canadian Library Association, 2000). The Canadian Guidelines are modeled on the ACRL Guidelines but are narrower in scope, reflecting issues and recommendations that are more appropriate to the Canadian context. The purpose of the guidelines is to emphasize the importance of planning and delivering effective library services to support distance and distributed learning programs.

The assumptions underpinning the Canadian Guidelines are articulated as follows:

(1) Access to library resources is essential for quality in postsecondary education regardless of where the learners and programs are located.

(2) Registered students and course instructors who are located away from the campuses of the originating institution are entitled to library and computer-based services as open and equitable as those provided for students and instructors on campus.

(3) Because distant learners are often disadvantaged in terms of library access, equitable library services in this context may involve more personalized services than would be expected on campus. It cannot be assumed that traditional library services, designed to support on-campus users, will meet the information needs of individuals involved in distance learning.

(4) The originating institution is responsible for ensuring that its distant learners have access to appropriate library resources including resources associated with the Internet and World Wide Web.

(5) Distant learners may choose to use local libraries for their academic needs, but if those libraries are unable or unwilling to provide the necessary support, the originating institution must be prepared to offer or arrange that support so the distant learner can acquire relevant resources or information.

(6) Distant learners lacking local access to relevant library resources or trained library staff require a means to obtain library materials and support services directly from the originating institution.

(7) Effective library support for a distance or distributed learning program requires advance planning by the Library in consultation with faculty, program administrators and other appropriate campus personnel, and with librarians at unaffiliated libraries.

The organization of the Canadian Guidelines is similar to that of the ACRL Guidelines, including sections on Definitions, Guideline Parameters, Philosophy, Finances, Administration, Personnel, Facilities, Resources, Services, Publicity, and Professional Development. The 1999 revision of the Guidelines recognizes the impact of developments in the information and communications technologies and distance education delivery methods on the provision of library services.

In the section on Resources, the Guidelines emphasize that:

Through either an unaffiliated library or a direct service from the originating institution, print and electronic resources should be made available to the distance or distributed learning program in appropriate number, scope, and format in order to:

- support the curriculum;
- assist distant learners in completing course assignments, projects, and theses;
- supply copies of recommended or supplementary readings or audio-visual materials to distant learners;
- support the library needs of faculty and instructors for course preparation or teaching;
- satisfy the need for reference and bibliographic information.

In the same section, the following statement delineates the role of the Library in providing access to electronic resources:

> With the advent of the Internet and the World Wide Web as well as the rapid proliferation of personal computers, resources are more widely available in an electronic format. Such electronic or digitized resources may include but not be limited to:
>
> - CD-ROM or online full-text, image, audio, video or audiovisual files;
> - descriptive or numeric datasets or databases accessible online or through electronic networks including the Internet/World Wide Web;
> - other online or networked resources such as FTP (file-transfer-protocol), listservs, NewsNets, the World Wide Web, and electronic chat rooms.
>
> As these electronic resources are more commonly accepted as legitimate resource materials, the Library should work to promote equitable access to these resources for the distant learner by means such as:
>
> - providing access to the appropriate computer technology and software at off-campus sites;
> - training users how to effectively locate electronic information on remote servers and how to download this information for local use;
> - instructing users in managing electronic information, including the appropriate style for citing electronic resources;
> - developing policies on adherence to copyright and appropriate computer usage;
> - supplying electronic documents in print format, in compliance with copyright law, to distant learners who are unable to access the appropriate technology.

Under Services, several points pertain to electronic resources:

> In order to meet the information needs of the distance or distributed learning program, a wide range of services may be necessary. All students and instructors in the distance or distributed learning program, regardless of location or country, should have a means to:
>
> - access, from remote locations, the online catalogue of the originating institution and any bibliographic or full-text databases which may be mounted on the system or otherwise available through the originating institution;
> - request the prompt delivery of library materials from the originating institution in cases where that material cannot be obtained easily and quickly through an unaffiliated library or by electronic means;
> - receive instruction or orientation in the use of libraries, library resources, or in automated library systems which are used for course-related research.

Implicit in the Canadian Guidelines is the assumption that the originating institution is responsible for all library services to its distance learners, both at home and abroad. As national boundaries are diminishing in distance learning, it is important that students located in other countries have the same or equitable access to library resources as do students located in the country of the originating institution.

GUIDELINES IN OTHER COUNTRIES

In Australia, the *Guidelines for Library Services to External Students* (Crocker, 1982) were developed in the early 1980s and have not been revised to date. Due to the geographic nature of this country, distance learning has been an integral part of Australian postsecondary education for many years. As a result, academic libraries in Australia have a long history of serving distance learners as part of their normal student body. Postal and document delivery services for distance learners have been in existence at many Australian academic libraries almost since the beginnings of the distance education programs (Crocker, 1991).

The Australian Guidelines reflect this long-standing tradition by emphasizing the logistics of how to support distance learners rather than the rationale and philosophy behind such services. The American and Canadian guidelines are more concerned with providing a document to justify the need for equitable library support at institutions that are inactive or minimally active in the provision of these services. The main areas of emphasis within the Australian Guidelines are the collection, access to library resources, information services, user guides, loans, copying, and charges. The Guidelines, of course, predate the electronic era and do not address issues that are of concern in today's world of online courses and virtual libraries. They are, however, still valid because there is a continuing necessity to maintain the traditional infrastructure for postal and document delivery services since electronic resources cannot satisfy the library needs of all distance learners.

In the United Kingdom, guidelines for distance learning library services do exist but are seldom cited in the literature. They are included as part of a Library Association publication titled *Library Services for Adult Continuing Education and Independent Learning: A Guide* (Fisher, 1988). The emphasis in this publication is on services to the adult learner from all types of libraries. Distance learners, including Open University students, are included in the scope of this document. Chapter 8 entitled "University Libraries and Adult Learners" is modeled after the 1982 ACRL Guidelines, providing a set of general guidelines targeted at the needs of distance and open learning students in the U.K. Major headings within the chapter include planning, finance, resources, services and facilities. These sections are followed by some more specific guidelines based on different types of courses and the different types of libraries within universities. The U.K. Guidelines appear to be little known and seldom used. This may be due to their inclusion in a more general document oriented towards the library needs of all adult learners.

There has been no indication that guidelines on distance learning library services are being developed in other parts of the world. Watson (1992) proposed a set of guidelines for developing countries but nothing further came of this proposal. It is interesting to note that individual leadership has been the driving force behind the creation of most sets of guidelines. While Watson's proposal was not backed by a library association, individual initiatives in other countries did receive support from the national library association. The guidelines for Canada and Australia were each developed primarily by one librarian working with a committee representing the national association. This indicates that the need for guidelines has been partly a bottom-up process, initiated by practitioners in the field who are concerned about quality assurance in distance learning. Apart from the United States, library associations have not been the main impetus for developing guidelines on distance learning library services.

In terms of international issues, use of new information and communications technologies is enabling universities and other educational providers to broaden their student base and expand their distance learning programs into other countries. The library implications of these ventures have not yet been fully addressed in any guidelines. The need for attention to this matter is likely to come from"victim" libraries when they feel the pressure of serving students enrolled at foreign institutions. The International Federation of Library Associations and Institutions (IFLA) may have a role to play in responding to these concerns. However, IFLA is unlikely to fill the need for guidelines since it has no history of developing documents of this type. As international programs become more commonplace, individual librarians again are likely to assume a leadership role in creating guidelines or adding sections to existing guidelines to ensure that the library needs of students in other countries are not neglected.

Some of the international issues that may need to be addressed in future guidelines include:

- Access to country-specific databases and library resources.
- How to provide offshore students with access to books and articles that are only available in print form.
- The need for formal arrangements with libraries in other countries to serve distance learners.

- Contracting with service providers such as the British Library to support distance learners.
- Provision of material in the native language of the country where the distance learners live.
- Provision of assistance and reference services to offshore students, especially those in radically different time zones.
- How to provide library instruction and information literacy training that is appropriate for the learning styles of students from other cultures.

QUALITY ASSURANCE AT INDIVIDUAL INSTITUTIONS

The library literature contains numerous references to the need for quality in off-campus and distance education programs. The subject index to the three annotated bibliographies lists close to 100 entries that discuss this topic (Slade & Kascus, 2000). In addition to works originating in the United States, authors from the following countries have contributed to the discussion: Canada, the United Kingdom, Australia, New Zealand, the Caribbean, and India. However, many of the comments on quality are anecdotal and merely point to library services as a means to enhance the caliber of distance learning. Case studies of individual institutions do provide some examples of successful practice in this regard.

One example is the University of Victoria in Canada. Seaborne (1997) applies the principles of quality assurance to the University's library service for distance education students. Even though the service does not have a formal quality assurance system in place, it demonstrates in practice many of the key elements which comprise such a system, including a service policy and plan, standards for service, procedures for service delivery, documentation of procedures, monitoring of service quality, and involvement of users. The Library's strategic plan, procedures manual, use of student evaluation forms, student surveys, and dedication of its staff contribute to the provision of quality services. It is pointed out that introduction of a more formal quality assurance system would have cost implications and would require a significant reorganization of the staff's roles and responsibilities in order to incorporate monitoring tasks.

As an illustration of the University of Victoria's approach to quality assurance, the following values are articulated in the strategic plan for the distance education library service:

1. Developing and maintaining credibility with all users by prompt and responsive service:
    *   requests for library materials from off-campus users and fee-paying clients are filled as quickly as possible;
    *   all inquiries for library information or materials are acknowledged given thorough attention;
    *   individual users are always informed of the status of their requests;
    *   requests from individual users take priority over other duties and tasks.

2. Taking a proactive approach to service:
    *   library needs are anticipated and mechanisms are developed to fill those needs as required;
    *   information and materials are collected in advance wherever possible.

3. Communicating and initiating dialogue with stakeholders:
    *   emphasis is placed on the exchange of information and cooperation with all relevant parties, both within and outside the University.

4. Promoting awareness of library resources amongst off-campus users and other interested parties:
    *   attention is given to conducting orientation and bibliographic instruction sessions and preparing promotional materials that emphasize the value of libraries for accessing and retrieving information.

5. Developing an information and research base in the area of off-campus library services and networking in the international library community:
    *   conducting research, publishing papers, and giving presentations on off-campus library services are regarded as significant activities to enhance both credibility and service delivery in this area;
    *   emphasis is given to sharing information and ideas about off-campus library services with colleagues at other institutions.

Another example is Deakin University in Australia. Deakin's approach to quality library services is described in a number of publications, the most recent by Cavanagh (2000). The University has earned a reputation for providing high quality library services to its off-campus students. The philosophy of library service for distance learners at Deakin University is that these students have the same rights to a quality library service as do on-campus students. Three broad types of services support Deakin's distance education programs: collection development, information services, and a library delivery service to all students, irrespective of their location and including some 700 overseas students. These services are complemented by the electronic resources available through remote access, a CD-ROM called the Deakin Learning Toolkit designed to provide students with the necessary software and instructions and tutorials for accessing and using the electronic resources, a reciprocal borrowing scheme in the state of Victoria, and a guide developed especially for off-campus students.

Deakin's Off Campus Library Service has developed a set of Performance Standards which ensure that students receive a consistent level of quality service. The standards are:

- 90% of book requests from off campus students are dispatched within 24 hours, if the item is available for loan.
- 80% of requests for photocopies of available articles are dispatched within 3 working days of receipt.
- 85% of subject/information requests are dispatched to students within 5 working days of receipt.
- 90% of material requested by off campus students is supplied.

The keys to the success of the Deakin's service are identified as: a willingness by the institution and the Library to support and fund the service, publicity and ease of access for students, a reliable method of delivering material, and an acceptable satisfaction rate for requests, as verified twice yearly in evaluation surveys of the Performance Standards.

A third example is the Open Polytechnic of New Zealand (TOPNZ), a distance education provider. The TOPNZ Library introduced a service quality charter in 1997 to improve services to library users (Clover, 1997). The charter was a project adopted as part of an overall service quality plan. It is written from the perspective of a library user and outlines service standards which users can expect to be met. Service quality is defined as "the extent of discrepancy between customers' expectations or desires and their perceptions." The participation of library staff in developing the service quality charter was a key element in the process. It resulted in customer service being placed high on the library's priorities, increased awareness amongst library staff of the service component of their work, and created a willingness to critically examine library procedures with the aim of removing barriers to customer service.

Excerpts from the TOPNZ Library Carter include:

Commitment to Customer Service:

- Customer service is the primary focus of the Library.
- We aim to assist you to make maximum use of the Library's collection and services.
- We aim to provide timely and relevant service and strive for consistent policies and procedures.

Communication:

- Staff will be available to provide help and advice between 8:30 AM and 5:00 PM Monday to Friday.
- Phone calls will be answered promptly in person or within five rings by automatic answering facilities.
- Outside working hours, messages can be left through voicemail, fax and email.

Services:

- We undertake to provide reference services to facilitate your successful use of library services, resources and collections.
- We will meet all reference requests within your requested time frame, or we will notify you of the reasons for the delay.
- We will action document supply requests within three working days and urgent requests within one working day.

One factor which all three institutions share in common with regard to service quality is support from their distance education providers. The libraries that are most successful in contributing to the quality of distance learning are recognized as an essential partner in the educational process by the departments or offices that develop and fund the programs. Without this recognition and support, it is unlikely that the libraries could offer or maintain the high level of services that they currently provide.

LIBRARY CONTRIBUTIONS TO DISTANCE LEARNING

The international literature of the middle and late 1990s shows a marked shift in emphasis from access to physical libraries and print materials to access to electronic libraries and electronic resources. With a growing number of remote users, whether they be distance learners or regular students and faculty, libraries need to develop services to support people who are accessing electronic resources from locations other than the campus libraries. The literature stresses that this support should be a team effort, involving partnerships with various stakeholders, the two primary ones being teaching faculty and computing services staff.

Martin (1998) indicates that librarians can contribute to distance education by forming partnerships with educators to design environments where information technologies can be integrated into the learning experience. For the provision of adequate library services to distance learners, Cajkler (1998) advocates that a librarian should be involved in all stages of course development.

This comment is particularly relevant since the author is a tutor of distance education courses rather than a librarian. Some of the ways in which librarians can support distance learning courses is by working with faculty to incorporate electronic resources into curriculum development, teaching students and faculty how to effectively use electronic resources, and generally attempting to foster information literacy skills in remote users. According to Miller (1997), the skills most needed by lifelong learners will be the ability to evaluate information, analyze it, and use it to solve problems. These are the information skills that librarians can foster in students as part of the new learning environment.

Zastrow (1997) indicates that librarians can assist faculty in designing and implementing electronic courses by educating them about current technologies, conducting literature searches, consulting on fair use and copyright, and giving input in designing user-friendly interfaces for computer mediated education. As an example of the contribution libraries can make to distance learning, Miller (1997) describes the development and evaluation of telematics applications at University College of Suffolk where library staff are compiling an online directory of subject or discipline-based web sites to offer students a "pre-validated" range of relevant sites for information and learning resources.

The issue of collaboration and forming alliances is emerging as a major theme in the literature. In addition to collaboration with faculty and course designers, another alliance deemed to be of importance in this context is one between librarians and computing services personnel (Haricombe, 1998). As libraries provide remote access to electronic resources for their users at a distance, it becomes important to offer technical support for that access (Hulshof, 1999). Collaboration with computing personnel can lead to effective support mechanisms for the remote user, with librarians handling the queries regarding information retrieval and search techniques, while computing services people deal with technical support and computer literacy issues. In addition to supporting distance learners, collaboration between libraries and computing services can benefit services to teaching faculty. For example, Jayne and Vander Meer (1997) suggest a new service role for librarians that involves collaborating with computing center personnel to teach faculty about instructional uses of the Web.

It should be noted that the works and examples cited in the preceding discussion on the role of the library are drawn primarily from developed countries like the United States, Australia, and the United Kingdom. In developing countries such as India, China, Africa, and the Caribbean, the literature on the role of libraries in distance learning has a different emphasis. The common theme in this body of literature is "barriers to service," with many authors emphasizing that library services for distance learners are underdeveloped and need to be improved. The issues most frequently discussed in this context include: the training of librarians and library staff, development of library collections at study centers and local libraries, expansion of physical facilities, access to basic equipment, the lack of adequate funding for library services, and, more recently, access to information and communications technologies.

ISSUES FOR UNIVERSITY AND LIBRARY ADMINISTRATORS

Regardless of the country or area of the world, libraries can contribute to the quality of distance learning in numerous ways. Heseltine (1995) predicts that, in the new learning environment, libraries will focus on actual user needs and develop more extensive end-user services that are not restricted by time or location. Emerging from this will be an organizational structure based on functionally oriented, collaborating teams in which individuals perform specific rather than generic roles in areas such as resource management, user support, document supply, network management, and external relations. MacDougall (1998) further predicts that, in the new academic environment, the librarian will become a hybrid manager utilizing professional knowledge, applying information technology, and contributing to curriculum design. Hahn and Li (1998) stress that librarians' skills and knowledge in using information technology will be of great benefit to distance learning programs, and academic libraries need to capitalize on their strengths in this area to support both distance learners and educators.

While librarians appear ready and willing to align themselves with faculty and administrators and participate in collaborating teams, there appears to be a general lack of understanding amongst faculty and teaching staff of how librarians can assist with instructional design and curriculum development. As a result, librarians are frequently left out of the distance education planning process and have to be assertive to make the stakeholders aware of the library's potential role in course development and delivery. Since library guidelines are aimed primarily at libraries, these guidelines may fail to impress faculty and administration or convince them of the need for library involvement in distance learning.

Research has shown that faculty and teaching staff tend to have minimal understanding of the concept of information literacy and how librarians can contribute to the development of these skills. A study at Southern Cross University in Australia investigated the attitudes and perceptions of teaching staff toward the information needs of their distance education students (Phelps, 1996). Interviews with distance educators revealed that many of them had confused and poorly formulated approaches to information skill development in their courses. While distance educators expected their students to access information resources, they did not necessarily expect the students to use library-based information. The author concluded that these approaches and expectations may be hindering the students' development of information literacy skills. It is emphasized that self-directed learning and information literacy are strongly connected and that teaching staff, librarians, and instructional designers have a major role in promoting these skills.

University administrators need to recognize the contributions that libraries can make to distance learning and provide the infrastructure and budget to enable librarians to fully participate in program development and delivery. They also need to encourage a climate that is conducive for librarians partnering with faculty and course developers. A campus-wide initiative to promote an under-standing of information literacy would greatly assist in this endeavor. Central Queensland University (CQU) in Australia is an example of an institution that has a commitment to helping its distance students develop information literacy and lifelong learning skills (Appleton & Orr, 1997). With support from the campus administration, the CQU Library has initiated a number of projects, in cooperation with teaching staff, to enable the students to access, evaluate, use, and manage resources electronically. Examples of these projects include: (1) a virtual residential school which used a combination of desktop video-conferencing and teleconferencing to teach students at a distance, (2) two computer assisted learning programs on using library resources, and (3) a series of courses on the Web to teach selected professional groups how to use the Internet effectively.

By partnering with faculty, librarians can help to create learning experiences that foster information literacy skills and the transformation of information into knowledge. In the developing information age, students graduating with such skills will have an advantage in the job market. The institution producing these graduates will also have a competitive edge in satisfying the needs and interests of its students and attracting future students.

Library administrators need to provide the resources and staff to ensure that their libraries can effectively support distance learning programs and information literacy training. Wolpert (1998) emphasizes that libraries should approach distance education as a new business opportunity using market evaluation and analysis techniques. If academic libraries align themselves closely with faculty and administrators and develop performance measures, they can provide educational support while improving awareness of the importance of libraries.

Some of the ways in which library administrators can contribute to quality assurance in support of distance learning include:

- Ensuring compliance with the ACRL *Guidelines for Distance Learning Library Services* or with other national guidelines as appropriate. In the absence of national guidelines, the ACRL and Canadian guidelines should serve as models.
- Actively promoting the concept of information literacy on and off-campus.
- Providing a standing invitation to faculty and teaching staff to participate with librarians in developing strategies for information literacy training and integrating library resources into distance learning courses.
- Developing partnerships with the campus computing services department to provide adequate technical support for library users who access electronic resources from off-campus locations.
- Creating and funding a library unit devoted to distance learning and remote access services.
- Providing a means for remote library users to obtain prompt assistance from a library staff member. This may include 24/7 reference service.
- Providing mechanisms for delivering library instruction to distance learning students. This may include tutorials accessible via the World Wide Web.
- Ensuring that all databases and electronic resources available to on-campus library users are also accessible to off-campus students in any part of the world.
- Conducting user studies to determine the library needs of off-campus and distance learning students and faculty.
- Conducting usability studies of electronic library resources to determine whether these products are meeting learning and teaching needs in the various distance learning programs.
- Providing a means for evaluating, on a on-going basis, the library services available to distance learners.
- Articulating service standards for supporting distance learners in a formal document such as a library charter or strategic plan.

Those libraries that can provide an effective support structure for distance learning programs, successfully market their services within the institution, and align themselves with faculty and administrators will be able to achieve a degree of recognition of their value in the new electronic environment. Those libraries that assume a less-active role may find that they lack the resources and finances to adequately serve their remote and distance learners. Under these circumstances, the distance learners may end up depending on electronic information without the assistance and training to critically evaluate and use this information. Ultimately, this may put the institution at a disadvantage in the global marketplace if its graduates are perceived as lacking essential information literacy skills.

CONCLUDING OBSERVATIONS AND FUTURE TRENDS

Librarians in most countries appear to be concerned about quality in distance learning. However, this concern manifests itself in different ways. Many individual institutions demonstrate aspects of quality assurance in the provision of library services to distance learning programs. One major indicator of quality assurance in library services for distance learning is the existence of national guidelines. The need for American guidelines has been driven largely by the Association of College and Research Libraries and the accreditation process. Outside of the United States, such guidelines seem to have been developed on an ad hoc basis rather than as a systematic initiative of a library association. These guidelines generally have been initiated by individual librarians from institutions that already demonstrate best practices in the provision of library services to distance learners.

Best practices from selective international institutions indicate that the most effective library support for distance learners involves a high level of support from both the campus and the library administration. In terms of future trends in the organizational nature of library services for distance learning, a distinction must be made between libraries that have a long-standing history of service to distance learners and those that are new to this area. Amongst libraries with established distance education services, it is likely that this function will receive even more support and recognition as part of an overall service plan to assist remote users. Authors such as Brophy (1997) predict that the library of the future will be a hybrid library where electronic and traditional services operate in tandem. This will be especially true in distance education services where users will continue to need access to print resources as well as electronic resources.

In institutions that are new to distance learning, a different pattern may emerge. Institutions without past experience in providing library support to distance learners may not be as sensitized to the unique characteristics of these individuals. There may be a tendency in developing new services to assume that remote access to electronic resources will meet most of their needs. Distance learners tend to have somewhat different characteristics and library needs than do full-time on-campus students. The average distance learner is an adult, often with work and family obligations, who is studying part-time. These learners generally lack peer group support for their studies, are often unable to access appropriate print resources through local libraries, and usually have very limited time for library research. In addition, the information retrieval skills of these students may be weak. The literature indicates that some of the newer educational providers are relying entirely on electronic products to satisfy the information needs of their students. This raises concern about a number of issues, including information literacy training, effective database and Internet searching techniques, limited access to information in print format, and the impact on local nonaffiliated libraries that are used by distance learners.

While the future trend in libraries with past experience in providing services for distance learning may be toward a hybrid model of delivery, the trend in newer institutions may well be toward a totally electronic support system. If executed with careful planning, a solid infrastructure, and adherence to national guidelines, these services could be effective. However, without proper attention to user education, point-of-need assistance, technical support, and access to non-electronic forms of information, these systems could contribute to less effective learning and lower educational standards. Of those institutions moving toward totally electronic services, the most successful will be those that perform the following functions: including librarians in the process of course development, offering a wide range of instruction and training in information skills to both students and faculty, and having a multi-faceted program of support and assistance for all remote users. These functions may be considered some of the future benchmarks for quality in the new online learning environment.

REFERENCES

Appleton, M., & Orr, D. (1997). Achieving Change at a Distance by Access to Information. In: S. Gregor & D. Oliver (Eds), *Processes of Community Change: Proceedings [of] a Colloquium Inspired by Professor John D. Smith, 31 October–1 November 1996* (pp. 131–137). Rockhampton, Queensland: Central Queensland University.

Association of College and Research Libraries. Distance Learning Section (2000). Guidelines for Distance Learning Library Services. *College & Research Libraries News, 61*(11), 1023–1029. Also online. Available: http://www.ala.org/acrl/guides/distlrng.html

Brophy, P. (1997). Off-Campus Library Services: A Model for the Future. *Journal of Library Services for Distance Education,* 1(1). Online. Available: http://www.westga.edu/~library /jlsde/vol1/1/PBrophy.html

Caballero, C., & Ingle, H. T. (1998). Thousands Still Shoeless: Developing Library Services in Support of Distance Education: A Case Study. *Journal of Library Services for Distance Education,* 1(2). Online. Available: http://www.westga.edu/~library/jlsde/vol1/2/ CCaballero_HIngle.html

Cajkler, W. (1998). Distance Learning: The Tutor's Experience. *Education Libraries Journal, 41*(1), 13–20.

Canadian Library Association. Library Services for Distance Learning Interest Group. (1993). Guidelines for Library Support of Distance Learning in Canada. CACUL Occasional Paper Series, no. 8. Ottawa: Canadian Library Association.

Canadian Library Association. Library Services for Distance Learning Interest Group. (2000). *Guidelines for Library Support of Distance and Distributed Learning in Canada.* Online. Available: http://uviclib.uvic.ca/dls/guidelines.html

Cavanagh, A. (2000). Providing Services and Information to the Dispersed Off-Campus Student: An Integrated Approach. In: *The Ninth Off-Campus Library Services Conference Proceedings: Portland, Oregon, April 26–28, 2000* (pp. 99–110), compiled by P. S. Thomas. Mount Pleasant, MI: Central Michigan University.

Clover, D. (1997). Committing to Customer Service: Development of a Service Charter at The Open Polytechnic Library. *New Zealand Libraries, 48*(12), 239–243.

Crocker, C. (1991). Off-Campus Library Services in Australia. *Library Trends, 39*(4), 495–513.

Crocker, Christine (Ed) (1982). *Guidelines for Library Services to External Students.* Ultimo, NSW: Library Association of Australia.

Dugan, R. E. (1997). Distance Education: Provider and Victim Libraries. *Journal of Academic Librarianship, 23*(4), 315–318.

Fisher, R. K. (1988). *Library Services for Adult Continuing Education and Independent Learning: A Guide. Library Association Pamphlet 40.* London: Library Association Publishing Ltd.

Gilmer, L. C. (1995). Accreditation of Off-Campus Library Services: Comparative Study of the Regional Accreditation Agencies. In: *The Seventh Off-Campus Library Services Conference Proceedings: San Diego, California, October 25–27, 1995* (pp. 101–110), compiled by C. J. Jacob. Mount Pleasant, MI: Central Michigan University.

Gupta, D. K. (1997). Library and Information Access to Distant Learners: New Opportunities Through Information Technology. In: A. L. Moorthy & P. B. Mangla (Eds), *Information Technology Applications in Academic Libraries in India with Emphasis on Network Services and Information Sharing: Papers Presented at the Fourth National Convention for Automation of Libraries in Education and Research (CALIBER–97) at Patiala, 6–8 March 1997* (pp. 40–42). Ahmedabad, India: Information and Library Network Centre.

Hahn, B. K., & Li, C. (1998). *The Role of the Academic Library in Distance Education.* Paper presented at the International Conference on New Missions of Academic Libraries in the 21st Century, Beijing, China, October 25–28, 1998. Online. Available: http://www.lib.pku.edu.cn/98conf/proceedings.htm

Hamlin, A. T. (1965). Library Services for College Extension Programs. In: U.S. Congress. Senate. Committee on Labor and Public Welfare. Subcommittee on Education. *Higher Education Act of 1965: Hearings ... on S.600. Part III* (pp. 1492–1506). 89th Cong., 1st sess. Washington, DC: U.S. Government Printing Office.

Haricombe, L. J. (1998). Users: Their Impact on Planning the Agile Library. In: L. J. Haricombe & T. J. Lusher. (Eds), *Creating the Agile Library: A Management Guide for Librarians* (pp. 81–93). Westport, CT: Greenwood Press.

Heseltine, R. (1995). End-user Services and the Electronic Library. *OCLC Newsletter, 215,* 19–21.

Holmberg, B.. (1989). *Theory and Practice of Distance Education.* London: Routledge.

Hulshof, R. (1999). Providing Services to Virtual Patrons. *Information Outlook, 3*(1), 20–23.

Jayne, E., & Vander Meer, P. (1997). The Library's Role in Academic Instructional Use of the World Wide Web. *Research Strategies, 15*(3), 123–150.

Latham, S., Slade, A. L., & Budnick, C. (1991). *Library Services for Off-campus and Distance Education: An Annotated Bibliography.* Ottawa: Canadian Library Association; London: Library Association Publishing Limited; Chicago: American Library Association.

MacDougall, A. (1998). Supporting Learners at a Distance. *Ariadne: the Web Version, 16.* Online. Available: http://www.ariadne.ac.uk/issue16/main/

Martin, R. R. (1998). The Library as a Partner in Creating a Dynamic Learning Community. Paper presented at the International Conference on New Missions of Academic Libraries in the 21st Century, Beijing, China, October 25–28, 1998. Online. Available: http://www.lib.pku.edu.cn/98conf/proceedings.htm

Miller, G. E. (1997). Distance Education and the Emerging Learning Environment. *Journal of Academic Librarianship*, *23*(4), 319–321.

Phelps, R. (1996). *Information Skills and the Distance Education Student: An Exploratory Study into the Approaches of Southern Cross University Distance Educators to the Information Needs of External Students*. M.D.Ed. thesis, Deakin University.

Roberts, J. M., & Keough, E. M. (Eds) (1995). *Why the Information Highway? Lessons from Open and Distance Learning*. Toronto: Trifolium Books.

Seaborne, K. (1997). Quality Assurance in the Provision of Library Services in British Columbia. In: A. Tait (Ed.), *Perspectives on Distance Education: Quality Assurance in Higher Education: Selected Case Studies* (pp. 77–88). Vancouver, BC: Commonwealth of Learning.

Slade, A. L., & Kascus, M. A. (2000). *Library Services for Open and Distance Learning: The Third Annotated Bibliography*. Englewood, CO: Libraries Unlimited.

Slade, A. L., & Kascus, M. A. (1996). *Library Services for Off-Campus and Distance Education: The Second Annotated Bibliography*. Englewood, CO: Libraries Unlimited.

Stephens, K. (1998). The Library Experiences of Postgraduate Distance Learning Students or Alice's Other Story. In: P. Brophy, S. Fisher & Z. Clarke (Eds), *Libraries Without Walls 2: The Delivery of Library Services to Distant Users* (pp. 122–142). London: Library Association Publishing.

Watson, E. F. (1992). *Library Services to Distance Learners: A Report to the Commonwealth of Learning*. Bridgetown, Barbados: Learning Resource Centre, University of the West Indies.

Wolpert, A. (1998). Services to Remote Users: Marketing the Library's Role. *Library Trends*, *47*(1), 21–41.

Zastrow, J. (1997). Going the Distance: Academic Librarians in the Virtual University In: *Computers in Libraries '97: Proceedings of the Twelfth Computers in Libraries Conference* (pp. 171–176), March 10–12, 1997. Arlington, Virginia, compiled by C. Nixon, H. Dengler & J. Yersak. Medford, NJ: Information Today, 1997.

TOTAL QUALITY MANAGEMENT: IMPLEMENTATION IN THREE COMMUNITY COLLEGE LIBRARIES AND/OR LEARNING RESOURCES CENTERS

Theresa S. Byrd

INTRODUCTION

By all accounts, the NBC white paper "If Japan Can, Why Can't We?" in the early 1980s was the event that started the interest in Total Quality Management (TQM) in America. Business and industry were the first organizations to utilize TQM, but by the mid-1980s higher education institutions and academic libraries/learning resources centers (LRCs) had joined the TQM bandwagon. Indeed, community colleges and academic libraries/LRCs that have adopted TQM believe that "It is a call to leadership for the reform of American enterprise. Its advocates want more than a change in management practice; they want an entirely new organization, one whose culture is quality-driven, customer-oriented, marked by teamwork, and avid about improvement." (Marchese, 1991, p. 5)

As the first tier of the American higher education system, community colleges extend educational opportunities to the masses of our society. However, in today's climate, where legislators and the public demand fiscal conservatism

Advances in Library Administration and Organization, Volume 18, pages 245–271.
Copyright © 2001 by Elsevier Science Ltd.
All rights of reproduction in any form reserved.
ISBN: 0-7623-0718-8

and quality higher education, community colleges are forced to do more with less. Community colleges, like all of higher education, are plagued by problems such as competition, budget constraints, decaying infrastructures, soaring tuition, shifting state and federal government priorities, and rapid technological developments. In the 1990s, as community college leaders ponder the future, they know that diverse student populations and distance education are additional issues with which they must deal.

In terms of adopting TQM practices, community colleges are the pioneers in higher education. Seymour and Collett (1991) report that "the most comprehensive TQM efforts are found at community colleges and smaller, private institutions" (p. 3). They note that community colleges see the use of TQM as a natural out-growth of teaching it.

This study examined the implementation of Total Quality Management (TQM) in three community colleges Library/Learning Resources Centers (LRC) or Instructional Support Services (ISS)[1] located respectively on the east coast; CC–EC, in the south, CC–S, and in the northeast, CC–NE. To place the LRCs' TQM experiences in proper context, the institutions' overall TQM experiences were also studied.

The purpose of this investigation was to determine how Total Quality Management has been implemented in community college libraries/LRCs and to describe ways in which it is assisting community college librarians and/or LRC personnel in attaining effective management, as well as improving their service-orientation, staff development, organizational structure, and utilization of technology to meet the demands of their communities. The goals of this study are to provide community college librarians and LRC personnel with information about how TQM as a management philosophy is being used in community college LRCs, and to aid them in determining if TQM is an effective management tool in terms of its ability to deal with organizational change and problems in community college LRCs.

This qualitative study, based on data from interviews, employs descriptive information to present the LRCs' TQM experiences in a real-life context. Moreover, a multiple-case replication design was used to provide a comparative analysis of TQM implementation in the three community college LRCs.

THE TWENTY FEATURE TOTAL QUALITY MANAGEMENT MODEL AND OTHER QUALITY CHARACTERISTICS AT THE THREE INSTITUTIONS

The Twenty Ideal Features of a TQM Initiative in Higher Education Institutions are stated below:

1. Identify a leader or champion to promote TQM in the institution. Demonstrate Commitment to and provide the leadership for implementing TQM.
2. Introduce the basic principles of TQM: philosophy, concepts, and tools.
3. Develop a plan with phases for implementing TQM. Such a plan should include mission, vision, values statements, and linkages with strategic planning.
4. Provide for a quality infrastructure to support the TQM initiative. This includes establishing a quality council or steering committee, appointing a quality coordinator, forming teams, and training personnel.
5. Implement an extensive training program for all college personnel that deals with job-specific and TQM matters.
6. Implement TQM in both the administrative and instructional operations.
7. Identify customers, both external and internal to the organization, know their needs, and commit to meeting those needs.
8. Select areas of quick and visible impact first to maintain momentum and enthusiasm in the early days.
9. Establish communication within the institution about TQM.
10. Promote a reward and recognition program to celebrate the efforts of teams and to sustain the TQM effort.
11. Use data to make decisions.
12. Reduce cost by attacking the processes, not the people.
13. Eliminate policies, rules, and regulations that hinder the progress of TQM.
14. Acknowledge that TQM must be integrated into the culture of an institution and not treated as an add-on.
15. Tailor TQM to the unique needs of each institution.
16. Understand that implementing TQM requires a considerable amount of time, training, and money, and that it is complicated and difficult to implement.
17. Expect some anxiety and cynicism from employees.
18. Realize that implementing TQM requires a change in culture – that is, an organizational transformation.
19. Remain focused on constancy of purpose, mindful that continuous improvement is a never-ending process.
20. Learn the benefits and pitfalls involved with implementing TQM.[2]

Despite the author's identification of twenty ideal features of TQM, it should be noted that the colleges in this study were successful in their pursuit of quality without endorsing all of the ideals. Nevertheless, using Walton's (1991) five stages for quality and the fact that organizations are reported to make the transformation to a quality culture in ten-year cycles, the researcher examined how CC–NE, CC–EC, and CC–S measured up against her twenty-feature TQM model.

DESCRIPTION OF THE THREE COMMUNITY COLLEGES' TQM INITATIVES

A Review of CC–NE's Stages of Quality

CC–NE in the tenth year of its TQM journey has approached Walton's (1991) fifth stage of TQM implementation. Stage one is the decision to adopt quality. In this stage, the leader of an organization accepts the premise that fundamental changes are necessary, then builds a consensus (Walton, p. 234). An administrator at CC–NE said that the decision to implement TQM at the college was made by the executive staff, and that the continuing support of the president was critical to the success of the effort. Other factors that influenced CC–NE to implement quality were an Assistant Dean of Instructional Support Services who was initially the primary advocate of adopting TQM at the institution, the airing of the NBC documentary, "If Japan Can, Why Can't We?" that introduced the concept of Quality through the Japanese experience to the American public, and the formation of the Philadelphia Area Council for Excellence (PACE), a Quality Roundtable of industry leaders of which CC–NE became a member.

Stage two is incubation. Having been converted to the principles of TQM, the leader takes an active role in promoting quality to overcome any doubt or hesitation about quality within the organization (Walton, p. 235). CC–NE, which calls its Quality process, Total Quality (TQ) has met this stage because its president and executive officers have taken an active role in promoting quality. The executive officers have experimented, along with others at the college, with a variety of quality governance structures. Currently, CC–NE has a twenty-five member Quality Council that consists of the executive officers, administrators, faculty, and support staff.

The CC–NE administrators and staff have read quality books such as W. Edward Deming's *Out of the Crisis*, Stephen Covey's *Seven Habits of Highly Effective People: Restoring the Character Ethic*, and Imai Masaaki's *Kaizen: The Key to Japan's Competitive Success*, and the institution's library has a quality collection. CC–NE's personnel have attended TQM conferences, participated in PACE and Covey training, sponsored training and retreats on TQM, and provided training to other organizations about quality. Additionally, CC–NE completed and distributed the Administrative Fundamental Notebooks to employees.

CC–NE's implementation plan relates to its mission, vision, and strategic plan. Indeed, CC–NE's personnel have taken steps to integrate the budget process, TQM, and strategic planning at the institution. CC–NE's president appointed a quality coordinator, and the union representative is involved in quality.

Stage three is planning and promotion. This is the stage where management and/or a quality coordinator develops a plan for introducing Quality to the organization, then promotes that plan in a visible way over a period of time (Walton, p. 236). CC–NE's quality coordinator plans and promotes TQM in the institution. The institution's promotion of strategies for making quality a part of its planning process include the college's linking of TQM with its budget and strategic planning process; Instructional Support Services' evaluation projects; the President's Recognition Award for support staff; and the institution's TQ training sessions.

Stage four is education. In this stage, management acquires an understanding of TQM principles and processes before asking others to serve on teams and work on processes (Walton, p. 237). CC–NE followed this dictate. Its executive officers and quality coordinator were the first to be educated about TQM process improvement and team skills. To assist in selling TQM to the college and to demonstrate how teams can be used to solve problems, CC–NE first organized three project teams: a copying team, a parking team, and a team that addressed placement of students in academic majors. Eventually, this number grew to a total of nine formal project teams. After its top managers were educated, CC–NE introduced other administrators and support staff personnel to TQM philosophy, tools, and processes. In fact, the ISS people indicated that each college department had engaged in writing its own mission statement, and that such statements were in addition to the college's mission statement.

Stage five is never-ending improvement (Walton, 1991). Stage five is characterized by the following:

(1) The quality coordinator, quality council, and facilitator have completed their work because now quality is part of everybody's job.
(2) The facilitators have been supplanted by supervisors, who have learned how to facilitate teams.
(3) The executive committee manages as a team, not an assemblage of private-fiefdoms.
(4) The annual "quality plan" and the "business plan" are the same document.
(5) The daily processes are monitored for trends and special causes that signal opportunities for improvement (Walton, pp. 238–239).

Of the five elements listed in stage five, CC–NE has not accomplished numbers one and two. CC–NE still has the need for the quality coordinator, quality council, and facilitators to ensure that Quality remains visible and is a part of everybody's job. And, although supervisors have learned how to improve processes and use TQM tools, they are not utilizing teams in their departments. They may, however, use TQM concepts in working with employees. Thus, there is still a need either for facilitators or for the institution to require supervisors to formalize the team process in their areas of supervision.

But CC–NE has mastered numbers three through five. Early in the institution's journey to Quality, the executive committee embraced Quality and began to manage as a team. The institution's strategic plan, *Vision 2000: A Work Plan for the Future* and the document, *Integrating TQM and Strategic Planning* confirm that the institution's "quality plan" and the "business plan" are indeed the same document. And, through its Administrative Fundamentals, CC–NE has begun to monitor daily processes for trends and special causes that signal opportunities for improvement.

Distinct Features of CC–NE's Quality Process
The fact that CC–NE has implemented TQM in a union environment is a unique feature of its Quality initiative. However, the most exceptional aspect of CC–NE's program is the progress that the institution has made with Quality governance by utilizing an iterative process that allows for changes in TQM coordination as people learn from experience what works and what does not. Another outstanding feature of CC–NE's Quality process is its linking of the budget process with TQM and strategic planning. It should also be noted that at CC–NE the institution has decided to meet the challenge of transforming its organizational structure by achieving horizontal integration of information and problem solving across all divisions.

A fourth exemplary feature of CC–NE's TQM experience is the way the institution instituted training for Quality. Also, the use of TQ mentors, the TQ support group for staff, the monthly lunch meetings for administrators to talk about Covey principles, and the vice president's round table where administrators discuss TQM principles, are special features of CC–NE's Quality effort. Additionally, the lessons that CC–NE learned from dealing with Quality over a decade is a distinct element of its journey to Quality.

Strengths and Weaknesses in CC–NE's Quality Process
As noted earlier, CC–NE's Quality process has evolved over a decade. The strengths of CC–NE's Quality initiative are these:

(1) progress in the area of quality governance;
(2) aligning TQM with the budget process and strategic planning;
(3) compiling Administrative Fundamentals notebooks;
(4) implementing horizontal integration;
(5) sponsoring TQ training;
(6) improving processes by teams;
(7) facilitating TQ administrator and support staff groups;
(8) the President's Recognition Award;
(9) identification of lessons learned;
(10) experience with the different stages of quality;
(11) offering quality courses;
(12) establishing ties with industry and the community.

The weaknesses of CC–NE's quality process are:

(1) that faculty were not required to participate in the TQM process;
(2) that with the exception of ISS, CC–NE's evaluation process still must be revised to make it more compatible with TQM by making the evaluation more team-oriented;
(3) that in high service areas with limited staffing, arrangements must still be made to allow staff to attend TQM and other training sessions.

Final Thoughts
CC–NE's attraction to Quality was galvanized because of its involvement with the Philadelphia Area Council for Excellence (PACE). At the same time, CC–NE's president implemented TQM to ensure the future success of the college. The institution's offering of Quality training, which generates revenue, may also have motivated the administrators to adopt Quality. In any case, CC–NE has made significant accomplishments in implementing TQM.

The fact that CC–NE has been involved with TQM for ten years confirms that the institution is committed to Quality. Both the administrators and ISS employees asserted that the college's TQ experience is still evolving and that the institution continues to seek ways to improve its Quality process. However, CC–NE's lack of inclusion of faculty directly in TQ weakens the systemic influence of TQM in the institution and allows an important group in the organization to ignore a cultural transformation that is underway. CC–NE administrators and ISS people, though, contend that, even if faculty do not label what they are doing as Quality, they are starting to participate in TQM through assessment and other activities.

A Review of CC–EC's Stages of Quality

CC–EC has approached Walton's (1991) fourth stage of TQM implementation in the third year of its TQM journey. CC–EC arrived at stage one, which is the decision to adopt Quality (Walton, p. 234), as a result of the president's relationship with the CEO at a local company. The CEO arranged and paid for a trip that allowed CC–EC's president to visit Jackson Community College to learn about how a community college was using TQM, and the CEO dispatched company Quality trainers to provide CC–EC's management staff with twenty hours of TQM training. Following the twenty hours of training, the CC–EC management staff made the decision to adopt TQM and to call its Quality process Building Quality Together (BQT).

Stage two is incubation (Walton, 1991, p. 235), which CC–EC reached when its mission, vision, strategic plan, and TQM processes became connected through its *Institutional Effectiveness Plan*. CC–EC also arranged for people to be trained by Fox Valley Technical College, sponsored speakers, trained facilitators, hired a Quality leader, established a Quality Leadership Council, instituted a Quality library, offered training to faculty and staff, and distributed Quality handbooks to all personnel.

Stage three involves the planning and promotion of TQM (Walton, 1991, p. 236). CC–EC's planning and promotion of TQM is demonstrated through each division's or department's plans as enumerated in their Planning Unit document and the Institutional Effectiveness Plans, which make Quality an integral part of the planning process at CC–EC. Additionally, at CC–EC, TQM is promoted through BQT training sessions and the recognition process for teams that complete a BQT project.

Stage four is education (Walton, 1991, p. 237), which involves management being fully trained in basic process improvement skills before introducing it to others. The institution has twelve BQT teams in operation, and the Quality office has conducted extensive training for employees. Indeed, the management staff also has been trained in TQM principles and tools. However, CC–EC only partially meets the requirements of this stage, because all of the LRC staff and at least one administrator indicated that upper management had not yet internalized the training.

Although CC–EC has teams working on Quality issues consonant with the college's goals, the institution has not reached the stage of never-ending improvement (Walton, 1991, p. 238), which is characterized by the following:

(1) The Quality coordinator, Quality council, and facilitator have completed their work because now Quality is part of everybody's job.

(2) The facilitators have been supplanted by supervisors, who have learned how to facilitate teams.
(3) The executive committee manages as a team, not an assemblage of private fiefdoms.
(4) The annual "Quality plan" and the "business plan" are the same document.
(5) The daily processes are monitored for trends and special causes that signal opportunities for improvement (Walton, 1991, p. 238).

Distinct Features of CC–EC's Quality Process

The most unique feature of CC–EC's Quality process is its decision to introduce quality into both the instructional and administrative processes of the institution at the same time. The development of the Continuous Learning Assures Students Success (CLASS) training modules and the use of faculty to train other faculty in CLASS principles further demonstrates the institution's commitment to connecting teaching and TQM.

Another outstanding CC–EC Quality achievement is that the institution serves as host to and provides the director for the Department of Community Colleges' Quality Consortium, which is dedicated to expanding and strengthening the use of continuous quality improvement concepts in community colleges statewide. Also, the LRC's staff involvement with TQM led to theirs and another LRC receiving a $210,000 library automation grant which was based on using the team approach to implement technology. In addition, two LRC librarians were trained as college TQM facilitators.

Other noteworthy features of CC–EC's Quality process include the development of linkages among its BQT process, budgeting, and strategic planning. CC–EC has developed a quality infrastructure, i.e., hired a quality leader, appointed a Quality Leadership Council, trained facilitators, and implemented teams. Remarkably, CC–EC has organized a systematic offering of BQT training sessions, at different levels, for all college employees. In addition, to provide the college community with the proper background to participate in quality processes, CC–EC developed and distributed the *Process Improvement: Teams and Tools Handbook*. The institution also distributed *The Memory Jogger: A Pocket Guide to Tools for Continuous Improvement* to all employees.

Laudable, too, is CC–EC's reward and recognition process that provides employees who complete a BQT project with a framed certificate and a $100 voucher to celebrate and display their team picture, with members' names, attached in the BQT Showcase in the main lobby of the Campus Center. CC–EC's rewards and recognition process is a step towards appropriately honoring the team rather than individual effort.

Strengths and Weaknesses in CC–EC's Quality Process
CC–EC's quality process, like that at any other institution, is characterized by strengths and weaknesses. The strengths of CC–EC's Quality initiative have been identified as:

(1) the introduction of Quality into both the instructional and administrative processes;
(2) the development of training programs and related handbooks and resources;
(3) the linkages among BQT, budgeting, and strategic planning;
(4) the improvement of processes by teams;
(5) the use of cross–training; the enhanced effectiveness of meetings;
(6) the recognition and rewards process; the offering of Quality courses;
(7) the establishment of ties with industry and the community.

The weaknesses of CC–EC's quality process are:

(1) the institution's failure to offer quality training quickly to new employees and to include all part–time employees in training sessions; and
(2) the perception that upper management is not committed to TQM because they do not "walk the talk."

Final Thoughts
CC–EC seems very involved in the process given the newness of its program. The fact that CC–EC has implemented TQM on the instructional side of the house indicates some level of seriousness about Quality. But, despite the fact that CC–EC teaches courses on quality, the institution only considered TQM after being prompted to explore it by the president of a local company. He arranged for CC–EC's president to visit a practicing TQM community college, and as a result, there are overtones that TQM was adopted at CC–EC for political and economic reasons, as well as to increase credibility for the courses on Quality it offered. Nevertheless, CC–EC's adoption of TQM has benefited its customers: business and industry, students, and the community. CC–EC is also recognized in its community college system as a leader in Quality.

A Review of CC–S's Stages of Quality

On a ten-year cycle, CC–S is in the infancy stage of its journey to Quality. Walton's (1991) stage one is the decision to adopt Quality (p. 234). With the president's decision to adopt TQM and his appointment of the Quality First Support Team (QFST), CC–S entered this stage of its Quality transformation. CC–S then moved to stage two, incubation (Walton, p. 235) by striving to incorporate Quality into the institution through the development of its mission and vision statements and the appointment of Work Out teams.

Stage three is the planning and promotion of Quality (Walton, 1991, p. 236), and the QFST is responsible for guiding and developing Quality at CC–S. The QFST is investigating different Quality programs in search of one that is appropriate for CC–S. Presently, the institution is considering adopting the Malcolm Baldridge Criteria for Education. In working with General Electric and Belmont University to receive training in TQM principles and processes, sending people to Covey and TQM training, as well as offering training for faculty and staff, CC–S meets stage four, which is education (Walton, p. 237).

Stage five is never-ending improvement (Walton, 1991, p. 238). Though CC–S displays evidence of cultural transformation through the downsizing of the administrative staff, the establishment of the Restructuring Team, the revised evaluation process, and the implementation of cross-training, the institution has not yet reached this stage. Stage five is characterized by these factors:

(1) The quality coordinator, quality council, and facilitator have completed their work because now quality is part of everybody's job.
(2) The facilitators have been supplanted by supervisors, who have learned how to facilitate teams.
(3) The executive committee manages as a team, not an assemblage of private fiefdoms.
(4) The annual "quality plan" and the "business plan" are the same document.
(5) The daily processes are monitored for trends and special causes that signal opportunities for improvement (Walton, p. 238).

Distinct Features of CC–S's Quality Process
CC–S's TQM process is unique for a number of reasons. CC–S is very successful and financially sound. However, CC–S, like many other institutions, has been forced to operate with more students and fewer dollars. CC–S chose TQM to help it find a way to make this adjustment. CC–S's administration expected TQM to assist the college in being more efficient and productive. Also, one administrator said that, with its TQM process, CC–S people came to understand that it was best to start a Quality initiative when the institution was healthy rather than wait until it was in decline.

There are aspects of CC–S's Quality process that are noteworthy. The fact that CC–S has no Quality coordinator makes its TQM process distinct. Many organizations and institutions have such a person. Yet, CC–S's president maintained that, if he named a Quality coordinator, people would believe that person was supposed to do all the Quality work. Thus, the president said to his constituency that, "no one is getting paid to do it, and everybody's getting paid to do it."

At CC–S, TQM and Covey training is voluntary. But it is remarkable that all three vice presidents and the president participated in Quality training, and that these people are now four of the 26 licensed facilitators who teach Quality classes and facilitate workshops at the college, in business and industry, and in the community. Moreover, CC–S has been most innovative in the revision of its evaluation process and in the establishment of a Restructuring Team.

Strengths and Weaknesses in CC–S's Quality Process
The discussion of the five quality stages and what makes CC–S's Quality process unique identified CC–S's strengths. They are:

(1) the implementation of TQM to gain efficiency;
(2) the involvement of the president and vice presidents in teaching quality courses and facilitating workshops at the college;
(3) the revision of the evaluation process to make it more compatible with TQM;
(4) the establishment of a restructuring team;
(5) the development of a staff training program;
(6) the offering of quality courses;
(7) ties with industry and the community.

There are also weaknesses, though, in CC–S's quality process. They are:

(1) the administrative and QFST employees' statements suggested that linkages have not been formed between TQM and the instructional process and academic assessment at CC–S;
(2) the LRC staff, with the exception of one person, disclosed that, initially insufficient time was spent on educating them about TQM, and, as a result, the college's emphasis on TQM had led to little change in their everyday lives.

Final Thoughts
Though CC–S does not have a Quality manual, everyone at CC–S received a Covey daytime planner. Moreover, in CC–S employees' Quality experiences, there are many familiar themes. Some of these are that:

(1) TQM is a fad;
(2) the institution embraced quality to transform its culture;
(3) the biggest barrier to TQM is faculty and staff resistance;
(4) TQM is not suitable for academe;

(5) students should not be called products;

(6) "the big word change scares people."

In reflecting about the CC–S quality experience, two observations come to mind. The CC–S administrative and QFST members were not impressed with the industrial model of TQM, and they indicated a desire to find a Quality process for CC–S that would not be too complex. On the other hand, the CC–S LRC staff seemed to be cynical about TQM. Their cynicism about Quality may have derived from the fact that these people believed their indoctrination into TQM implied that they did not do Quality work, and they resented the implication of being non-Quality and not customer-oriented.

Moreover, while CC–S may have adopted TQM to assist the institution in transforming its culture to ensure its competitiveness, it is also true that the offering of Quality courses and other training is financially lucrative for the institution. Many at CC–S asked why the president decided to embark on Quality three years before his retirement. An administrator shared that the president embraced Quality to strengthen the institutional culture to enable the institution to, not only survive the changes that will come with new leadership, but influence the future of the college as well. But that being said, the future of TQM at CC–S is unclear with the president's retirement. There are those at the institution, though, who hope that TQM will continue to thrive.

THE COMPARATIVE ANALYSIS OF ADMINISTRATORS AND LRC PERSONNEL RESPONSES

Administrators

Introduction. At CC–NE, CC–EC, and CC–S all the administrators were very positive about TQM implementation efforts in their colleges, and all were knowledgeable about TQM. In all three cases, a local business assisted their institutions to become involved in Quality. An administrator in each institution mentioned contacting Fox Valley Technical Community College for assistance in beginning their Quality initiative. The administrators agreed that a goal for their institutions in pursuing TQM was to improve efficiency in how their colleges operated.

Effective management. The administrators stated that at their institutions linkages existed between the TQM implementation plan and the mission, vision, and strategic plan. The administrators stated that each of their institutions had a Quality council to govern its TQM implementation processes. Likewise, the administrators indicated that all of their institutions were using teams to improve processes. CC–S did not have a Quality coordinator, but both CC–NE and CC–EC believed the Quality coordinator position was central in their initiatives.

The administrators at all of these institutions stated that Deming's philosophy served as the foundation for their colleges' Quality efforts but that their institutions also incorporated the ideas of other Quality gurus into their TQM implementation processes. All the administrators indicated that their institutions used TQM tools. However, CC–NE and CC–EC seemed to have more experience using such tools than CC–S.

Administrators in the three institutions approached the problem of dealing with employees who do not commit to Quality in slightly different ways. CC–S chose the approach of listening to resistors and valuing differences in thought. CC–NE decided to treat dissenters gently and patiently because top level administrators believed that this was the best approach to preserve the institution's family culture and because those administrators believed that people embrace Quality at their own pace. CC–EC, on the other hand, chose to deal with noncommittal employees by involving them via teams in improving processes.

Administrators at both CC–EC and CC–NE admitted that there was a lack of congruency between TQM philosophy and performance evaluations in their colleges. At CC–NE, though, one administrator mentioned that the ISS Division had piloted an evaluation process based on TQM principles. Conversely, CC–S achieved congruency between TQM philosophy and performance evaluations at the college by switching from criteria and checklists to performance agreements for everyone in the institution. CC–NE and CC–EC have established mechanisms to reward and recognize teams, but CC–S has not dealt with this issue to date.

In responding to whether TQM offered the college something that other management tools did not, the administrators indicated that it did. A CC–NE administrator said that prior to TQM, he learned by trial and error but now his actions were based on theory. Two CC–EC administrators said that TQM helped them look systematically at processes in a way that they had not done previously. A CC–S administrator said that the idea of using teams to improve processes was what attracted his institution to TQM.

Service to customers. From the administrators' remarks, CC–EC is the only one of the three institutions to implement TQM on the instructional side of the college. However, both CC–EC and CC–NE administrators acknowledged that there is a link between TQM and assessment in their institutions. CC–S administrators said that their institution's Quality effort does not include instruction or a link between TQM and assessment. An administrator at CC–NE and one at CC–EC perceived a connection between TQM and regional accreditation process[3] All the administrators confirmed that their colleges offered TQM courses for business and industry.

Staff development. The administrators at each of the institutions stated that their college provided TQM training for employees. At CC–NE, though, because of the union, faculty were not required to participate in training. The administrators indicated that their institutions spent a considerable amount of money on TQM training. For example, CC–EC spent an estimated $20,000 and CC–S spent approximately $40,000 a year on TQM training for the year studied. No figures were available for TQM training at CC–NE, because the institution included such training in its regular staff development budget.

Organizational restructuring. With the exception of establishing a quality infrastructure and eliminating middle management positions through attrition, none of the administrators at the three institutions provided evidence that their institutions had implemented radical organizational change in terms of moving from a vertical to a horizontal organizational structure. On the other hand, CC–NE administrators, through horizontal integration, and CC–S administrators, via a Restructuring Team, are working to create new organizational structures in their institutions.

Future of TQM. When asked what is unique about implementing TQM in a community college, a CC–NE administrator and a CC–S administrator both maintained that the TQM implementation process was the same for all institutions, but a CC–EC administrator said that community colleges' relationships with business and industry made the implementation of TQM unique in these types of institutions. Continuing with the TQM implementation theme, the administrators at CC–NE, CC–EC, and CC–S explored ways to take TQM to the next level in their organizations. For instance, administrators at the three institutions recognized the potential value of using the Malcolm Baldrige National Quality Award Education Pilot Criteria. Additionally, the administrators at these three institutions were searching for ways to institutionalize TQM into the culture of their schools.

Though the administrators mentioned a variety of benefits that their institutions received from implementing TQM, they all acknowledged that a specific benefit of TQM was better interpersonal relationships — that is, people working together on teams and seeing each other as customers and suppliers. While the administrators at the three institutions identified many obstacles to TQM implementation in their institutions, they all cited people's attitudes and resistance to change as an impediment to establishing TQM in their organizations. The majority of the administrators across the three colleges stated that some progress has been made in dealing with these barriers but not all of the roadblocks to TQM implementation in their institutions had been overcome.

At each of the institutions one or more administrators admitted that not everyone in the college was on board the TQM bandwagon. The administrators realized that they were not perfect in providing leadership for implementing TQM or changing the institution's culture from a bureaucratic structure to one based on a Quality philosophy, but their impressions of their commitment to TQM and how they "walked the talk" was stronger than that expressed by the LRC personnel.

LRC Personnel

Introduction. The feelings of the people in CC–NE's ISS Division regarding the TQM process can be described as positive. Overall, the CC–EC's LRC employees were positive about the implementation of TQM, though two or three staff members did express negative, doubting opinions about the initiative. However, CC–S's LRC personnel, with the exception of one individual, were negative and cynical about the implementation of TQM. The staff in the ISS and the LRCs said their organizations' adoption of TQM was due to larger collegewide Quality commitments. The ISS or LRCs also adopted their college's quality name as the label for their quality processes. As for the ISS's and LRCs' goals in adopting TQM, a librarian in each of the three organizations either stated specifically or implied that providing better service to customers was the reason for engaging in Quality.

Personnel in CC–NE's ISS, CC–S's LRC, and CC–EC's LRC differed in their opinions about whether TQM threatened the library's or LRC's traditions, standards, and respected bodies of knowledge. Both CC–S and CC–EC people believed that TQM threatened the traditional framework of the library or LRC profession, especially for staff members who preferred rules and standards over providing service to customers. Conversely, CC–NE people said that TQM did not threaten the traditional benchmarks of the profession, but these individuals also indicated that TQM helped the library to improve.

Effective management. ISS and LRC employees pondered if their departments' TQM implementation plans related to their missions, visions, and strategic plans. Of the three organizations, the CC–EC LRC which had a purpose statement; the BQT planning unit document, which lists goals, objectives, and strategies for a one–year period; and the college's budgeting process connecting with TQM came closest to fulfilling all four elements. Still, the CC–EC LRC did not have a vision statement.

All ISS personnel at CC–NE said that their individual department's TQM implementation plan related to its mission, vision, and strategic plan. Perhaps these individuals meant that, in theory, there was such a close alignment between their department's TQM implementation plan and the four elements of TQM because of college work in these areas. The reality, though, is that the departments have only mission statements and Administrative Fundamentals, a handout that contains the mission statement, customers, and flowcharted key processes. None of the individuals in the ISS mentioned the formal existence of a department vision statement, a strategic plan, or a TQM implementation plan.

CC–S has only a mission statement. Moreover, it differs from CC–NE and CC–EC, because the majority of the LRC individuals said that they were not aware that the LRC had a TQM implementation plan, and they did not mention the existence of a vision statement or a strategic plan.

As for how involved the ISS's or the LRC's top management employees were in the TQM process, the CC–NE's library director, CC–EC LRC's dean and department heads in both organizations stated that they were very involved in the implementation of Total Quality in Learning Resources. However, at CC–S, only one librarian said that the LRC's top management employees were very involved in the TQM process. The remainder of the staff expressed difficulty in being able to gauge the involvement of the LRC's management in Quality in the LRC or the institution.

A comparison of the ISS and the LRCs in developing a council or team to guide the TQM process revealed that CC–EC with its Learning Resources Council was the most advanced in this area. Though a librarian at both CC–S and CC–NE said that the whole staff was a team, the reality is that in these two organizations the forum for bringing the staff together is a staff meeting; a TQM team or council does not exist. As for TQM tools, staff in all three organizations stated that they used flowcharting and brainstorming. In addition, the CC–EC and CC–NE staff said that they used the Plan-Do-Check-Act (PDCA) cycle. The CC–EC and CC–NE staff made extensive use of TQM tools, but the majority of the CC–S individuals indicated that they were neither knowledgeable about nor utilized TQM tools.

The staff in the three organizations expressed two viewpoints about the impact of TQM on workload. In both the ISS Division and the LRCs, one group of individuals contended that TQM did create more work because of the time involved to do Quality training, cross-training, TQ projects, paperwork, and the planning of meetings. Another group of individuals, however, maintained that in general workload increased in their operations, but they did not see a connection between TQM and the increase in workload. These individuals cited the reason for the increase in workload as more customers, increased demands, or an automation project.

A CC–EC department head and the library director and two department heads from CC–NE agreed that, managerially, TQM made them value the viewpoints of the people on their staff. Also, these individuals viewed themselves as coaches for or in partnership with their staff. The prevalent belief of the CC–S department heads was that TQM resulted in more work being passed on to them; which, consequently led to them becoming overloaded.

The ISS staff and the staff in the LRCs identified benefits related to the implementation of TQM in their organizations as cross–training and the improvement of services to customers. These individuals perceived barriers to TQM implementation to be that change is difficult and that people are reluctant to change. It was the perception of staff in the ISS and the LRCs that the barriers related to effective management had not been overcome.

Service to customers. An analysis of the ISS's and LRCs' staff comments about the ability of TQM to improve services to patrons revealed that CC–NE and CC–EC personnel unanimously agreed that TQM did help improve service to customers, and the CC–S people, based on their limited experience with TQM, said that it could help improve service to customers. The staff of the ISS and LRCs said that customer service survey data provided feedback from customers about their satisfaction or dissatisfaction with LRC services. In addition to surveys, the CC–S and CC–NE staff said that they used suggestion boxes to solicit responses about services from students and faculty. Each of the three organizations provided evidence of how they used customer feedback to improve LRC services. For example, CC–S and CC–EC decided to purchase a library automation system and to improve copier service, and CC–EC instituted a no-food policy in the LRC based on problems identified by survey respondents.

A CC–S librarian, the CC–NE staff, and the CC–EC staff affirmed that there were similarities between their traditional service attitudes and the customer focus of TQM. Even so, the majority of the employees in the three organizations believed that TQM resulted in the staff being more focused on the desires and needs of customers. ISS and LRC staff members in the three organizations also contended that, with or without TQM, the customers were their focus.

Both ISS and LRC staffs cited the enhancement of customer service in Learning Resources and an increase in recognition and support by people on campus for providing good service as benefits of TQM related to service and customers. But there was no agreement among the ISS and LRC personnel about the barriers to TQM and service in their organizations. At CC–S a department head listed unreasonable service requests by students as a barrier to TQM and service. A department head at CC–EC commented that traditional thinking and old paradigms were barriers to TQM in the service area. And, a CC–NE department head asserted that the failure of the college to deal with TQ in hiring people meant that new employees might not have a commitment to a strong customer service philosophy. As for whether or not their organizations had overcome the barriers related to TQM and service, the ISS staff and the people in the LRCs responded in one of three ways, saying: (1) that they could not identify a barrier related to TQM and service, (2) that some of the barriers to TQM and service had been removed, or (3) that none of the barriers to TQM and service had been removed.

Staff development. The employees in ISS and the LRCs said that their institutions provided TQM training for college employees, which included LRC personnel. Both CC–S and CC–NE personnel stated that their institutions also provided Covey training for employees. A CC–EC person acknowledged that Covey could assist a college in assessing its personnel's commitment to TQM, and she stated that the Carolina Quality Consortium had undergone three days of Covey training. Among the three organizations, all individuals on staff at the time their colleges offered TQM training were trained, but new employees were not. The ISS Division and the LRCs had a critical mass of people trained in TQM principles because the ratio of new people was no more than two or three out of eleven at CC–S and CC–NE and two out of eleven at CC–EC.[4]

The LRC personnel at CC–EC and the ISS personnel at CC–NE all agreed that TQM had changed their staff development programs. At CC–S, though, only one out of four respondents believed that TQM had changed the LRC's staff development program. Management personnel in all three organizations indicated that TQM made them realize the importance of encouraging staff to develop by receiving training and taking courses. Only the CC–EC LRC sponsored training through the Learning Resources Council for the LRC staff. At CC–S and CC–NE, staff development training was handled by the institution rather than the LRC. ISS and LRC personnel were active in Quality at the institutional level. At CC–EC, the LRC Dean and a department head served as BQT training facilitators, and, at CC–NE, a department head served as both a TQM and Covey trainer. The ISS's staff and LRCs' staff stated that their colleges offered staff development training on topics such as customer service, the Internet, voice mail, and WordPerfect for Windows. However, the majority of the focus group members at CC–EC and CC–NE did not believe that TQM helped them do their jobs better.

All of the CC–EC and CC–NE employees listed benefits related to TQM and staff development, but three of the four respondents at CC–S indicated no benefits in this category. But despite the ambivalence of CC–S's staff on this issue, the benefits derived from TQM and perceived by the staff in the three organizations were: (1) that more staff development was offered by the college for employees, (2) that staff development helped the staff come together as a team, and (3) that training assisted the staff in doing their jobs better.

The barriers to TQM and staff development among the ISS and LRC personnel centered around problems relating to finding the money and time for training while still maintaining services, especially given staff scheduling problems. Of the four managers responding from the three organizations, only one CC–NE department head indicated that her department, by setting aside a half-day a month for training, has taken steps to overcome these barriers to TQM and staff development.

Organizational restructuring. Individuals in the three organizations cited several elements of TQM that had an impact on the ISS's and LRC's organizational structures. These included cross-training, teams, downsizing as well as changes in job descriptions, changes in the organizational charts, and development in service areas. However, none of the ISS or LRC employees reported that TQM significantly changed their organizations' traditional, hierarchical structure. Whether it is through CC–NE's library's weekly librarians' meetings, CC–S's LRC's staff meetings, or CC–EC's Learning Resources Council meetings, all three organizations seemed to be making strides towards consensus decision-making. Yet some employees in all three operations believed that the decision-making process had not changed enough, because top management still exercised too much control. Furthermore, the composition of CC–EC's Learning Resources Council and CC–NE's weekly librarians-only meetings indicated that other staff may not be very involved in the decision-making process.

The ISS and LRC staffs perceived benefits related to TQM and the organizational structure in the three organizations to be that everyone on staff became a stakeholder under TQM and that TQM allowed everyone to participate in what the LRC did through the use of teams and projects. But, the ISS's and LRCs' personnel said that TQM, if done properly, takes a lot of time, and this commitment served as a barrier in the transformation of the organization's structure. A second barrier mentioned by supervisors in both CC–S and CC–EC was that staff do not want to accept the responsibility that comes with empowerment. These barriers are significant in that it was the opinion of the majority of the staff in the three organizations that the barriers they enumerated had not been overcome.

Technology. In either the ISS or the LRCs, the replacement of the card catalog with an automated library system or an upgrade of an old automated library system to a new one was the way in which the staff in the three organizations saw a connection between TQM and how their operation would meet the technological challenges of the information age. Also, a CC–NE department head and the CC–EC staff said that technological changes required employees to receive more training about both computers and software in order to meet customers' needs.

As for benefits related to TQM and technology, CC–NE's ISS staff and CC–S's and CC–EC's LRC personnel indicated that TQM helped them get funding from either a grant or the college to purchase a library automation system. The CC–NE and CC–EC staff acknowledged that it is important for everyone to be trained in how to use the new technology. An individual in CC–S, CC–NE, and CC–EC mentioned that the librarians' fear of technology, either because of change and learning and working with new equipment or because they believed that technology would replace them in their jobs, was a barrier that related to both TQM and technology. The respondents in the ISS Division and the LRCs indicated that they were making progress in overcoming some of the barriers related to the relationship between the implementation of a TQM program and technology.

CONCLUSION

Administrators

A review of administrators' statements revealed that CC–NE's long history with Quality has allowed the institution to gain experience with everything from a variety of Quality councils to horizontal integration. Therefore, CC–NE is more advanced than either CC–EC or CC–S in the areas of Quality governance and in the development of a supportive organizational structure. CC–EC, from the very beginning of its Quality process, decided to implement TQM simultaneously in both the administrative and instructional sides of the institution. In addition, CC–EC required faculty to participate in TQM training and to use it in the classroom. CC–EC's mandatory training program for faculty to receive TQM training is noteworthy in comparison with CC–S's and CC–NE's volunteer TQM faculty training programs. CC–S, though, excels in the area of evaluation. Of the three institutions, only CC–S has changed its evaluation plan college wide to be more in keeping with the TQM philosophy.

Advice to other community college administrators. The administrators at the three institutions expressed common experiences about lessons learned from implementing TQM. In fact, an administrator in each of the institutions advised colleagues in implementing TQM to be patient and tolerant. It is not easy to change an organization, and implementation takes time. A CC–S administrator and a CC–NE administrator also cautioned colleagues to remember throughout the TQM implementation stages that Quality requires all the groups operating in an organization to be included in the process.

LRC Personnel
Of the ISS and LRC organizations, CC–S's LRC was impacted least by TQM. Although the CC–S LRC staff was trained in TQM principles, these individuals were confused about what it was. They were particularly confused about the difference between Covey and TQM principles. These individuals did not use teams nor did they improve processes with TQM. In fact, the use of TQM in CC–S's LRC was very elementary or nonexistent.

The CC–NE's ISS staff made moderate use of TQM. All the administrative personnel were very knowledgeable about TQM; and the focus group members were somewhat knowledgeable about TQM. Both groups were aware of the difference between TQM and Covey. Moreover, they had improved processes via their Administrative Fundamentals and the evaluation process. The ISS's employees' flaw in implementing TQM was that they did not use formal TQM teams and the professional librarians were excluded from the TQM experience.

At CC–EC's LRC, TQM had a significant impact on the organization. The majority of the staff understood TQM philosophy and expressed a commitment to Quality. The CC–EC LRC personnel had formed teams and used teams to improve processes. These LRC people also established a Learning Resources Council to provide leadership for their organization, and they held retreats which focused on how to improve the implementation of TQM in Learning Resources.

Advice to other librarians or LRC Personnel. An examination of CC–NE's ISS people's remarks and CC–S's and CC–EC's LRC employees' comments regarding what they would do differently if they could change the current TQM process revealed a common theme of "involvement" among the personnel in the three organizations. At CC–S, LRC people suggested that a college must involve everyone in TQM and keep people informed about the TQM process as well as avoid distinctions between Covey or TQM trainers and other campus employees. CC–EC's personnel, in considering what they would do differently, advocated that a college must include a mixture of people in training sessions, especially part-time people and upper management. A CC–NE person advised that, to guarantee the continued involvement of all personnel in TQM, a college should have people periodically review TQM principles.

The ISS and LRC personnel considered management models and whether they would select TQM again. The CC–EC staff firmly stated that they would select TQM again, and CC–NE's staff's comments were positive as well. By contrast, the CC–S staff did not respond affirmatively about selecting TQM again. One person, though, expressed interest in continuing with the Covey principle rather than TQM.

The three staffs agreed that they would inform fellow colleagues thinking about implementing quality of four things – first, that TQM takes time and work; second, that the institution's and LRC's top management employees must support TQM; third, that top managers at the institutional level and in the LRC must relinquish control in favor of participatory management for TQM to work; and fourth, that the TQM effort must involve everyone in the institution.

Also, the staff of all three organizations emphasized that it was important for colleagues to understand that they must invest in training to ensure that people understand the TQM process, to change people's attitudes, and to provide people with information about how to work together on teams. CC–S and CC–EC LRC employees also advise their colleagues elsewhere that those who choose this course must advertise the Quality effort widely on campus, and that such publicity should include information about both the successes and failures of TQM.

As in the TQM literature, the staff in these organizations supported both sides of the argument as to whether or not TQM can be implemented in an isolated area or whether it must be implemented college wide. A CC–EC person said that TQM must be implemented college wide or not at all, while a CC–S person said that it was possible to implement TQM in a single department. In all three organizations, it was clear from the perspectives of people both in the ISS Division and personnel in the LRCs that not everyone embraces TQM.

Areas of TQM to be Strengthened in the LRCs

Each of the three schools, regardless of how long they have been involved in TQM, has made good progress in developing a Quality infrastructure and in doing preliminary work toward institutionalizing TQM in their organization's culture. However, for TQM implementation efforts to be strengthened, the administrators and ISS and LRC personnel in the three institutions need to concentrate on making improvements in four areas: decision making, TQM training, organizational restructuring, and commitment to TQM.

Decision making. Administrators and leaders in ISS and LRCs must learn to rely more on a participatory, consensus management style. In the case of

Instructional Support Services and the Learning Resources Centers, the leadership must understand that their employees do not just want to come to meetings and talk or be informed about decisions after they are made. Staff, in keeping with the tenets of TQM, want to participate in the decision-making process. At the same time, administrators at the start of TQM must inform staff that participatory management does not mean that employees make all the decisions.

TQM training. Administrators and ISS and LRC staff indicated that TQM training was not offered continuously. This means that at the colleges and in ISS and the LRCs new and/or part-time employees are not trained in TQM principles. It is difficult for new and/or part-time employees to develop a Quality mentality or assimilate into a Quality culture when they have not been trained in TQM principles. The lack of commitment to TQM principles that is displayed by untrained new and/or part-time employees is revealed by an LRC person who said: "I've been in meetings and had to pick up all the little things that they do. I can shake my head while they're doing it." The failure of the LRC to train this person resulted in his showing little appreciation for TQM, and he even appeared to mock the process.

Organizational restructuring. Neither administrators nor the ISS or LRC personnel indicated that the institution or the ISS or LRCs operated in other than a hierarchical manner. Perhaps the ISS and LRCs cannot revise their organizational structures until their colleges tackle the problem of restructuring. Indeed, the CC–EC LRC staff at their 1996 summer retreat did discuss making changes in its organizational structure, but the staff decided that such change would not work so long as the college used a rigid personnel classification system.

Commitment to TQM. Administrators and ISS and LRC personnel believed that both the college's and ISS's and LRCs' leadership could work on improving their commitments to TQM.

Transformational Effects of TQM in the LRCs

Based on CC–NE's Instructional Support Services', CC–S's Learning Resources Center's, and CC–EC's Learning Resources Center's initial TQM experiences, there is some evidence that TQM has the power to transform libraries and Learning Resources Centers. In this study, the transformational effects of TQM on the ISS and LRCs occurred in the following areas:

(1) developing a mission and strategic plan,
(2) attaining effective management, implementing cross-training,
(3) offering staff development,
(4) using technology,
(5) conducting effective meetings.

The transformational effects of TQM had the least impact when the organization was trying to utilize consensus in decision-making, develop participatory decision making styles and restructure the organization. In addition, all three organizations had overlaid the implementation of TQM on a bureaucratic organization and violated the principle that TQM is not an add-on but a cultural transformation and the development of a new organizational structure. However, this should not be surprising, given that Coate (1990), Entin (1994), and Torbet (1992) stated that it takes five to ten years for Quality to take hold in an institution.

From the ISS and LRC employees' experiences, it is possible for other LRC personnel to understand the successes and pitfalls of TQM implementation. As set in the context of their institutions, the Quality experiences of the ISS and LRCs illustrated that the TQM implementation approach is important in terms of the systemic effect of quality on an organization. Also, it is clear that a union versus a nonunion environment affects an institution's or LRC's TQM implementation plan and its TQM progress, especially in terms of the involvement of faculty rank personnel in the initiative.

The three case studies revealed that the LRC's leadership role is significant in promoting with the staff the goal of never-ending improvement and in maintaining the momentum of the quality effort. At CC–NE and CC–EC, the ISS and LRC employees served as TQM or Covey trainers in the college, an indication that such involvement assisted them in developing their organizations' TQM initiatives. This is certainly true in comparison with CC–S's LRC in which TQM was found to be nonexistent and where the staff complained about not being involved in TQM at the institutional level. CC–NE's ISS and the CC–EC's LRC were recognized by administrators at their respective institution as one of the units on campus which had embraced TQM.

Most importantly, the remarks of administrators and ISS and LRC personnel indicated that institutional and LRC leaders must consistently "walk the talk" in their commitment to TQM or others in the LRC and the institution will become disillusioned. The good news, though, is that many of the LRC employees believed that, if the president, the deans, and the LRC leadership make a commitment to Total Quality Management, the philosophy can flourish in institutions of higher education and in libraries and Learning Resources Centers that support them.

NOTES

1. CCNE's Instructional Support Services contains the same units as the Learning Resources Center in CC–S and CC–EC.
2. This model was developed by the author based on extensive reading of TQM literature.
3. CC–NE – Middle States Association of Colleges and Schools and CC–EC Southern Association of Colleges and Schools. in their institutions.
4. CC–EC has seventeen people on the LRC staff but only eleven participated in this research and ten of these eleven people were trained in TQM principles.

REFERENCES

Brassard, M. (Ed.) (1988). *The memory jogger: A Pocket Guide of Tools for Continuous Improvement.* Methuen, MA: GOAL/QPC.
Coate, L. E. (1990, November). TQM on campus: Implementing Total Quality Management in a University Setting. *NACUBO Business Officer, 24.5,* 26–35.
Cornesky, R. A., Andrew, R., & Merrick, W. (1990). *W. Edwards Deming: Improving Quality in Colleges and Universities.* Madison, WI: Magna Publications.
Cortada, J. W., & Woods, J. A. *The McGraw-Hill Encyclopedia of Quality Terms & Concepts.* New York, NY: McGraw-Hill, 1995.
Covey, S. R. (1990). *The Seven Habits of Highly Effective People: Restoring the Character Ethic.* New York, NY: Simon and Schuster.
Deming, W. E. (1986). *Out of the crisis.* Cambridge, MA: Massachusetts Institute of Technology, Center for Advanced Engineering Study.
Entin, D. H. (1994, May). Whither TQM? A second look: TQM in Ten Boston-area Colleges, One Year Later. *AAHE Bulletin, 46.9,* 3–7.
Imai, M. (1986). *Kaizen: The key to Japan's Competitive Success.* New York, NY: Random House Business Division.
Jurow, S., & Barnard, S. B. (1993). Introduction: TQM Fundamentals and Overview of Contents. *Journal of Library Administration, 18.1/2,* 4.
Lewis, Ralph G., & Smith, D. H. (1994). Total Quality in Higher Education. Delray Beach, FL: St. Lucie.
Malcolm Baldrige national quality award education pilot program. (1995). Gaithersburg, MD: National Institute of Standards and Technology.
Marchese, T. (1991, November). TQM Reaches the Academy. *AAHE Bulletin, 44.3,* 3–9.
Seymour, D. (1993). *On Q: Causing Quality in Higher Education.* Phoenix, AZ: Oryx Press.
Seymour, D., & Collett, C. (1991). *Total Quality Management in Higher Education: A Critical Assessment.* Methuen, MA: Goal/QPC.
Torbert, W. R. (1992, December). The True Challenge of Generating Continual Quality Improvement. *Journal of Management Inquiry, 1.4,* 331–336.
Walton, M. (1991). *Deming Management at Work: Six Successful Companies that use the Quality Principles of World-Famous W. Edward Deming.* New York, NY: Perigee Books.

APPENDIX

Continuous improvement – uses specific methods and measurements to systematically collect and analyze data for the purpose of improving the processes identified as critical to the organization's mission. The components of continuous improvement are both a philosophy and a set of graphical problem-solving tools or techniques, such as brainstorming, the use of flowcharts, Pareto charts, control charts, and scatter diagrams (Jurow and Barnard, 1993, p. 4).

Customer – In the academy, students, alumni, and employers share some characteristics of traditional customers. A student is a customer while enrolled in a university but he become a product of the university once graduated. A further parallel to the customer concept lies in the relationships among various components of the university. For example, virtually every operating unit is a "customer" of the maintenance department (Cornesky, Andrew, and Merrick, 1990, pp. i and iii).

Malcolm Baldrige National Quality Award (Baldrige Award) – Established in 1987 through legislation (P. L. 100–107), the award is named for Malcolm Baldrige, who served as Secretary of Commerce from 1981 until his death in 1987. Both businesses and educational institutions use the Baldrige criteria. However, because of the uniqueness of educational institutions, in 1993 a decision was reached to launch pilot criteria for education based on the 1995 criteria for the Award (*Malcolm Baldrige National Quality Award Education Pilot Program 1995, p. 1 and inside cover*).

Total Quality Management – TQM is founded on the understanding that organizations are systems with processes that have the purpose of serving customers. TQM calls for the integration of all organizational activities to achieve the goal of serving customers. It seeks to impose standards, achieve efficiencies, to define roles of individuals within process and the organization as a whole, to reduce errors and defects by applying statistical process control, and to employ teams to plan and execute processes more efficiently. It requires leaders willing to create a culture in which people define their roles in terms of being responsible teammates and in terms of the values they add in delivering quality outputs to customers (Lewis and Smith, 1994, p. 10).

PAPERS FROM THE DOUGHERTY SYMPOSIUM AT THE UNIVERSITY OF MICHIGAN NOVEMBER 1999

The University of Michigan held a symposium in honor of former Library Director Dr. Richard M. Dougherty, November 1999, in recognition of his distinguished career and many contributions to the library profession. Four papers were commissioned from prominent faculty and librarians. These papers were to address some of the most compelling issues facing libraries and the scholarly community during the next several years. We are pleased to offer them here to make them available more broadly to the profession.

As library director, Dick Dougherty relentlessly worked to improve operations while advancing the profession and mentoring countless colleagues. He was especially interested in the innovative, those ideas and changes that reached beyond the status quo to improve services and advance the profession. He shared his enthusiastic perspective through editorials in his Journal of Academic Librarianship, and through his professional engagement, including during his term as President of the American Library Association. This symposium paid tribute to his many accomplishments and contributions to the profession. The papers that follow capture that same sense of forward-looking vision to reach beyond current practice and cause us to think about innovative initiatives that will lead us into the decades ahead.

William A. Gosling
Director
University Library, University of Michigan

Advances in Library Administration and Organization, Volume 18, page 273.
Copyright © 2001 by Elsevier Science Ltd.
All rights of reproduction in any form reserved.
ISBN: 0-7623-0718-8

CHANGES IN SCHOLARSHIP AND THE ACADEMY AND, PERFORCE, ACADEMIC LIBRARIES

Paul N. Courant

CHANGES IN SCHOLARSHIP AND THE ACADEMY – INTRODUCTORY MUSINGS

I was asked by Bill Gosling to discuss the subject of "scholarship and the academy," with emphasis on their relationship to leadership, change and risk-taking in academic libraries. The terms "scholarship" and "the academy" invite many possible interpretations, so I start by defining them as I use them in this discussion.

I think of scholarship as the craft of learning and teaching. The word "scholarship", by the way, is used much more in the humanities than in the sciences and social sciences, and its use generally carries this connotation of craft. Classicists and art historians talk about the quality of scholarship in terms of command of the literature, elegance and completeness of sources consulted and sources used (both primary and secondary). An argument is incomplete and unpersuasive if the scholarship is not good, even if the rhetoric and logic are first rate. In the natural and social sciences there is much more discussion of the quality of research measured by laboratory or statistical technique. But it's the same stuff. It is the craft of learning and persuasive conveying of what one has learned – hence of teaching as well as learning.

Advances in Library Administration and Organization, Volume 18, pages 275–286
Copyright © 2001 by Elsevier Science Ltd.
ISBN: 0-7623-0718-8

And that's what we professors do, I think. We learn, and we convey our learning (and the techniques of how we learn and how to convey learning persuasively). I think I can fit most anything I might want to say about scholarship under that definition. People may want to argue with me about the connotation of good craftwork, but there is in my view nothing more satisfying than a job done well, so if they want to argue about that connotation they should not do so because I am talking about "mere" craft. There is nothing mere about it – it is an exalted craft. I also think that this definition works pretty well from before the invention of movable type to after the invention of the web. There are surely changes in our craft and how we practice it, but it's been about learning and teaching all along.

With learning and teaching come judging. Scholars spend an enormous amount of time evaluating other scholars and would-be scholars. We act as referees for each other's work for publication, hiring, and promotion. Discipline by discipline we generally agree on what is meant by "good" work; where we don't, we have tremendous battles that sometimes spawn new disciplines or clusters of disciplines. It is entirely natural, I think, that the craft of learning and teaching brings with it the ability and inclination to judge others' learning and teaching. I also think that this capacity and inclination to judge and hence to certify the quality of others' work, the broad set of things that we call "peer review", are at the heart of our ability to make a living as scholars. And, to get ahead of my story, the fundamental processes of teaching, learning and evaluating others' teaching and learning seem to me to be unchanged by the new technology, or even by old ones. These are the things that we do and will continue to do.

The academy is the place, or set of places, where we sit and do our work. When it's the place, it's like the sea as the place where mariners ply their trade, it is an idea, not a geographic location. There are also specific academic institutions, thousands of them. I'll limit my discussion to research universities and their immediate kin, but one of the points I want to make (which has been made by many others) is that new information technologies may have significantly different effects on the academy and on "typical" academic institutions, such as Michigan or Emory or Harvard or the Santa Fe Institute. We have been referring to the "academy" as a virtual location far longer than we have been using the word "virtual" with its current connotations. We think of it as a sector of the intellectual world and a sector of the economy. We also think of it as a set of particular institutions. The first meaning isn't going to go away. In contrast, some have argued that the set of institutions that currently comprise the academy, and the types of institutions within that set, are going to look radically different pretty soon. (And I find, as I think these issues through, that this is the most interesting question – will individual universities as dense physical agglomerations of scholarly activity survive, and what will they look like if they do? I like universities, they are where I live and work, and also I find the question interesting in its own right. I got my start as an urban economist, and most everything in urban economics flows from trying to understand the distribution of human activity in space.)

The academy – indeed the individual academic institution – is of course the home of the academic library, in no small part because (at least many) scholars require libraries in order to do their work. Meanwhile, librarians, lovers of books and archives, producers of scholarship and vital enablers of others' scholarship, spend an endearing amount of time and psychic energy worrying about where the next meal, or at least next acquisition, might be coming from. In an economy and a set of institutions that are increasingly fee for service or implicit fee for virtual service, it's pretty scary to be in a line of work where on a good day you might sell things with a market value that approaches 1% of your costs.

But budget comes last, and it should. One of my current job titles is Associate Provost for Academic and Budgetary Affairs. Scholarship and its support are academic affairs, arguably THE academic affairs. The purpose of budget is to help make the good stuff happen, and to sharpen the choices that are inevitable. But leadership requires that what one wants before getting ground down by what one can and cannot have. So the order of the topic that Bill Gosling assigned to me is the right order – first scholarship, and then the academy, and last of all, the academy as an entity that must balance its books. And that will be the order of the remainder of this discussion – how have changes in scholarship changed the workings of the academy, and, by the way, what might this have to do with the way that we do library budgets, just to pick an example. (This neat ordering won't quite work, because some of the changes in the academy actually do not derive from changes in scholarship, but from other relevant forces.)

CHANGES IN SCHOLARSHIP

For most of us, the new technologies are still in some combination a better filing cabinet, a better encyclopedia, a better post office, a better copy center, a better phone system, a better tape-recorder, a better camera, and yes, a better library, but they haven't really changed how we go about doing things. Note, however, how many of the things I listed are themselves fairly new: Copy center, phone system, camera (at least video), tape recorder – even typewriter. That's how revolutions appear when you are living through them. A new thing comes along, and it's different from the old but performs an old function (we call it the type WRITER). Then another new thing comes along, and after a lifetime or two, the shape of some domain of human activity is really very different.

I had occasion to reflect on this recently when I bought my wife a new camera, a pretty good one. It does all of the things that my fifty-year-old Leica does, and of course it obeys all of the same rules of optics. You aim it by looking at the subject and you push a button to take the picture. And yet it's almost a different instrument. It has dozens of program modes. It asks intelligent questions of its operator and proposes answers that are usually right. It provides an enormous amount of information about lighting, focus, and its own frames of reference (program modes) for thinking about these matters. The set of skills that I developed 40 years ago that allow me to operate the old Leica aren't necessary to operating the new Canon (although the basic knowledge about what cameras do is still the same). Moreover, there are a host of things that have to be learned on the new camera that are entirely irrelevant to the old Leica. But they are both cameras, both of high optical quality, and indeed I can take the same picture with both of them and the result will be essentially indistinguishable. How much change do we have here? It depends on how we measure it. Is this a revolution? Well, yes and no, or maybe and maybe not.

I would argue that the piece of new technology that has caused the most change in general scholarly practice so far has been the e-mail attachment. (To verify this, read David Lodge's wonderful comedy of the academy, Small World, copyrighted in 1984, just before ubiquitous e-mail. Lodge's characterizations of academic life are still funny and still true, yet the absence of e-mail clearly dates the book as a period piece.) Even the klutziest group of academics can now write joint papers without having to arrange meetings (if on the same campus) or scheduled phone conversations punctuated by return mail or return FedX (if at more than one site.) So the technical problems in producing the standard form of collaboration – the multi-authored book or paper – are easily solved for pretty much any pair or trio of scholars, at any plausible set of scholars' locations in the world. That's actually a big deal. It is surely the case that elaborate collaboratories involving real-time exchange of large volumes of information will come to be increasingly important, and that desktop video will come to enhance our current means of exchanging documents. Still, except in a few fields, it's going to be a long time before scholars do much more than attach files to messages, secure in the expectation that their collaborators will work the files over and send them back with messages of their own.

This change has two flavors of impact on the academy, one salutary and one troublesome, both inevitable. The salutary change is that it is easier to collaborate, to exchange information and ideas. When graduate students leave, they can easily continue to work on projects that they started with mentors and other students. Collaborations can better survive the exigencies of tenure denial, outside offer, and peripatetic spouses. New collaborations can form easily, requiring relatively little in the way of face-to-face contact – perhaps only a few hours spent together at the annual meeting of the relevant professional organization. All of that is terrific, except that it weakens faculty attachment to the place where they happen to work.

The new technology makes it easy for me to work with colleagues at Stanford and Berkeley. Thus I am less interested than I used to be in my department's recruiting in my field or related ones. I don't mean to say that I am uninterested, merely less so. And less so is enough to make it harder to make departments and universities run well, to induce the myriad activities that faculty have traditionally engaged in that keep the university running. In the old days, such activities had a much stronger investment component than they do now – they were essential to creating the environment in which one would do ones' work, and hence were necessary to being able to do one's work well. Now, less of the scholarly environment is local, so there is less payoff to making these investments. This, I think, is one of the reasons that we see increasing pressure to pay faculty (usually in release time) for administrative work and other service burdens. The cost of not doing such work was once felt directly, at least to some degree, as an inferior academic environment, and also as annoyance on the part of one's colleagues for failure to pull one's weight. Now, with the academic environment less tied to the scholar's location, that cost is diminished. We have not figured out how to deal with the problem, to restore the value of local institution building. I am skeptical that we will ever deal with it fully, because the new technologies have changed the constraints facing both faculties and their institutions. (And I don't mean to pretend that the good old days were golden. I merely claim that their silver tarnished more slowly and was somewhat easier to keep clean). This may also be the cause of what seems to me a surprising demand for increasingly high-quality space. Because people can work at home and on the road, institutions strive to improve the environment on campus. (I'd love to figure out how to test this last proposition.)

My colleague Michael Cohen, a big fan of new technologies, would point out about now that jet aircraft may be even more important than e-mail. It was jet aircraft that first made it plausible that faculty from Stanford and Harvard, or Harvard and the other Cambridge, could collaborate relatively easily, even if it involved getting together for a week or two. Whether it's Eudora or Boeing as cause, the fact is that the development of national and international communities of scholarship makes it harder to run local ones, and leads local ones to compete economically by purchasing services from faculty that were once part of a more informal economy. And there is nothing we can do about the causes, which will continue to weaken the local and continue to lead faculty to demand (and in a competitive market receive) explicit reward for providing services that help the institution but not the individual career.

Some observers of the academy and information technologies have argued that "soon" (that wonderful infotech word that means anything from "it's going to happen this afternoon" to "I know someone who was talking to some folks at MIT and there's this really cool thing . . .") faculty attachment to physical academic institutions will be like that of independent contractors. That's possible, I suppose, but it's a long way from where we are, and I will be surprised if the trend away from faculty concern with local environments will go anywhere near that far. In principle, the technology will enable such a world. But in principle, the printed book enabled the end of lecture courses, and it didn't happen that way, for very good reasons. Note that in this area, as in many others, the sciences got there first. Over a century ago, they invented the workshop as a collaborative mechanism. Steamships and steam engines were adequate if a group was going to spend many weeks together, which is exactly what happened in those old scientific congresses. As we shall see, it is also the scientists who have made the most complete use of the new technologies in the practice of their craft, although they still like to get together for weeks at a time in nice places.

RESULTS FROM AN INFORMAL SURVEY

The biggest change in scholarship that I really know is my own. I've gone from empirical economics, where I would analyze and draw inferences from large data sets, to a method that I call administrator's ethnography, where I try to figure out what is going on in various domains of the university by asking a few people who might know something. (I still use data, but rare is the day when reliable data gets me very far towards knowing what to do, although it still often lets me rule out really idiotic possibilities.) When I was called upon to prepare this talk I employed this new method, and sent e-mail to a few dozen colleagues around the country, asking them how their scholarship had changed over their careers, as well as to reflect on the broader topic that I was to discuss In other words, following conventional administrative practice, I asked them to write my talk. They were pretty helpful. Answers to my questions were all over the map, with students of information focusing changes in the academy and others focusing on changes in scholarship.

John King, who recently became Dean of our School of Information here at the University of Michigan, sees a revolution in the academy. Following a cogent analysis of what it is that we sell to the world, and how it is that we got to become such a valued enterprise in American life, he suggests that in the future, much of our teaching activity will be distributed out from the academy, because the technology allows it. I'm sure he is right that there will be much use of distributed learning and teaching, but it remains to be seen whether this will radically change the shape of, say, the University of Michigan. Undergraduates, whose coming together in one place is core to our coming together in one place, like to come together. There are substantial agglomeration economies to their growing up, to their education, and (as John Seely Brown and Paul Duguid point out) to the really important part of their education, which is their learning how knowledge is produced in groups by people who understand each other.[1] Students learn from each other, about many things, and many of those things require the high bandwidth of face to face and hand to hand contact. All of this leads me to believe that at least for the middle class and above, the residential undergraduate institution is going be alive and well for a long time.

Given that there are still going to be clumps of students, the logic of agglomeration economies suggests that faculty should be nearby. Moreover, there are also agglomeration economies in research, and in having research take place in the same neighborhoods as teaching, notwithstanding e-mail attachments and virtual laboratories. The obvious connection is that research and teaching feed each other – they are both parts of scholarship. (This is a topic that I could go on about at length. Are research and teaching substitutes or complements, competitive or synergistic? Yes, is the short answer, although it's fun to argue both sides.) Additionally, people like to talk with each other and often profit by doing so. Although people may be less attached to the clusters they are in than was once the case, they need a lot of hardware and support to do their work, and the resulting scale economies are themselves enough to support agglomeration. (Note that the big research labs that aren't parts of universities are big and getting bigger, as are the big research labs that are parts of universities.) And, once all of these scholars are in the same neighborhoods, those neighborhoods will generally be the sites where people come together to figure out to distribute the increasing fraction of scholarship that will be distributed.

Although many very cool things with new technology are happening in the arts and the humanites – new media as well as greatly enhanced ability to manipulate and display artifacts with digital technology, my informal census of colleagues unequivocally indicates that the biggest changes in everyday scholarly practice are in the sciences. Physicists get up in the morning and look at a website that contains essentially all that anyone is publishing in the areas that they care about. As one of my colleagues in physics puts it:

> Now when one finishes a paper it is posted on the (Los Alamos) archive. The next day anyone in the world who signed up gets a copy of the abstract, and can print the paper if they want, or anyone can sign on the archive and see what papers were put there, read them, print them, etc. Publication and peer review is no longer relevant to the research, and we no longer read or subscribe to journals. We still publish in journals because we have to have a way to prove to our Chairs and Deans that we do research (since there is a pretence of merit raises), and often we collaborate with students and postdocs who need publications in refereed journals to get jobs, but all the publishing is for bureaucratic reasons. Research now is like a real time dialogue among everyone in the world who is capable. Someone makes a little step, we all learn of it, integrate it and take the next step, always with time lags of days or weeks. We no longer save the information, simply relocating it whenever we need it. It's a wonderful process, very suited to how people function, too, and most of us love it. We also seldom have conference proceedings anymore, simply posting the transparencies on a conference site, so anyone can go there and look at them or print them. It's almost as if everyone in the world who works on things I'm interested in is in the same place – not quite since personal conversations over coffee or wine still have a major role to play, but otherwise similar.

Now, that's a pretty different world than the one I grew up in, and also from the one that most humanists and social scientists live in. By the way, eventually that world will solve the problem of outrageous publication costs for scientific publishing (a subject that Richard Dougherty has had smart things to say about for a long time) but that will still be a long time in coming. Note, by the way, that the world described by my friend only works in communities of discourse that are tight enough so that refereeing as such is relatively unimportant to judging whether work is good or not. That's easiest in disciplines that are very strongly unified in methods and practice, and it is these disciplines, I think, that are able to make the best use of the new technologies for general scholarly communication.

Here are comments from another scientist, focusing on a different kind of use of shared information:

> It is now possible for me to access, within seconds, the entire database of gene sequences, and I can use powerful programs available on the web to look for genes or proteins with similarity to the one I am studying. In addition, I can easily pull up any one of more than 5000 protein structures, examining the overall structure and the locations of individual components on my desktop computer. So my ability to relate the proteins I am studying to others that have been studied is incredibly enlarged. The result is a rapid development of protein phylogeny, where we look at the properties of large and distantly related groups of proteins and seek to understand how their function and structure evolved.

This is closer to the changes in my own field. I can access huge amounts of census data, and also spatial data in various Geographical Information Systems, suitable for analysis and mapping using programs and techniques that are easily available to people working in the area. And I can go back and get refreshed data tomorrow or next month. And my colleagues who are good at this kind of stuff can actually automate the updating process to some degree. These are big differences in the way we do our work, wildly different from doing it by hand, less wildly different than ordering a tape that used to come through the mail, except that now I can assign the material to students, point it out to colleagues, etc., much more effectively. Note that my examples from economics and the example from protein phylogeny have little to do with the effect of information technology on the social and communication structure of scholarly activity, they are more about the computer as a tool of scholarly inquiry – a better file-drawer cum calculator. This should not be forgotten. When they got their start, computers were called computers because we used them for computation. Our ability to analyze data, simulate complexity, etc. is still growing, and that growth still matters, and matters a great deal. (And it's another reason for clumping of research. One needs to be well wired to a lot of processing power.)

LIBRARIES

Individual academic institutions, with lots of faculty and students, are going to exist for a long time in part because the payoff to propinquity will continue to be high for a long time. Nothing persuades me of this more than Michigan's own experience with collaboratories. The idea behind collaboratories is to allow people at multiple sites to work on the same set of problems, problems that require lots of data, including real-time experimental data. Our School of Information is a pioneer in this area. Our School of Information is also (like every other unit on campus) adamant that they cannot be an effective academic unit effective unless they have lots of high-quality contiguous space that allows faculty to interact with each other and with students face-to-face. It's going to be a long time before propinquity stops mattering, and that may be enough to hold universities and their libraries together.

More generally, there is my experience in space planning. Absolutely everyone is willing to pay top dollar (and even more willing to have the central administration pay top dollar) for high-quality space near where everything else is. This tells me that on the research side, as well as in teaching, it is still the case that there are agglomeration economies. And some people even want to be near the library. In both the "old" academy and the "new," libraries are institutions that help glue places together – they generate agglomeration economies.

Given that there will still be places that are academic institutions, what will be the role of libraries? Libraries will still perform the function of helping scholars to get easy access to the information they need. The changes in this function, I opine, look more significant to librarians than they do to the scholars.

Of course, physical archives will continue to be important and which will continue to be a reason for agglomeration. But Wendy Lougee and others point out to me that more and more of what the library is called upon to do is to know where things are, that are not in the library. This is a tougher problem than knowing where things are, that are in the library, although even the latter problem isn't trivial. I'm dying to know how librarians are going to solve the problem. But as a scholar, I really don't care whether the library gives me a book or a journal or a URL.

I also want something more, something that I believe is increasingly hard to deliver. I want to have all sorts of rich and reliable cues about how good the information in the location might be. The new technologies make this problem harder than it used to be, in two important ways, both of which I want librarians to solve.

The first way involves something might be called bibliographic integrity. The standard bibliographical citation to a book or a journal is a triumph of metadata – it tells us exactly where to go to find the information that is being used. Just to be super-sure, the old rules called for giving the city of publication, as well as publisher and title, so we would know exactly what was being referred to. Moreover, I could trust my friendly neighborhood academic library either to have or to find the source I wanted to check upon. Even for rare books, the network of libraries would eventually produce what was required, unambiguously. URLs have none of these nice properties. Even the official sites maintained by governments change frequently. As a general matter, websites come and go and their content changes all the time. So I may cite a version of a paper attached to someone's homepage, and next week when you look to verify my claim you will find a different version, or nothing at all! It could be that the author and owner of the website will be careful about this, and clearly label different versions, but nothing in the system requires it, and it is the extreme exception, rather than the rule. I want librarians to figure out how to return us to a world of reliable and replicable scholarly citation. How are librarians are going to do that? How are the rest of us to go about our business if they don't?

The second problem involves separating the wheat from the chaff in an environment where a search engine generates thousands of hits. How am I supposed to know which of the many sites cited are any good? Michael Cohen has pointed out to me that the overall effect of the quantity of information available online has been to take to a new scale the scholar's problem of allocating extremely scarce time across very large numbers of potentially interesting possibilities. The paradoxical consequence of this is – and increasingly will be – to make us more reliant on disciplines – and, somewhat equivalently, on old-boy/girl networks – for clues on what to attend to, increasing their influence. I would like librarians to mediate this process so that we are not so reliant on what we learned and who we met in graduate school for figuring out what work is worth reading. The new technologies open up the possibility of greatly increased mixing and learning across the disciplines, yet if Cohen is right, and I think he is, they also make us more reliant on routines of quality judgment that work, and the routine that works best is highly disciplinary. Just how libraries and librarians can organize to vet the quality of information is an open question. But if libraries don't do it, I fear that some combination of the disciplines and popularity will, and I don't like those outcomes.

CONCLUSION

From my outside perspective the basic functions of libraries in the academy are unchanged. Tell me where the information is, or better yet (as Dougherty was fond of putting it when he ran the library) make it appear in front of me. Assure me that if I look in the same place as I looked yesterday (or last year, or last century) I'll find the same thing as I found then.

If libraries keep performing those functions, provosts will keep funding them (assuming that provosts continue to exist, which I think they will). Of course, the funding will always be difficult because libraries are classics of a phenomenon in economics known as a public good, something that cannot recover its costs efficiently through charging fees for service. Adam Smith understood perfectly well that public goods must be provided through extra-market mechanisms, generally via governments. In the university context, the government is the central administration, which one hopes will be responsive to the interests of faculty and students. But I see nothing new here. Libraries have always been public goods. What is sort of new is that for much content it will eventually come to be the case that the "system" needs only two copies, an original and a backup, on independent servers. We are no where near figuring out the organizational arrangements to make things work this way – we are too used to having one of everything nearby. The public goods problem is always harder to solve the bigger is the compass of the relevant public, and so this one will be harder to solve. But we will solve it, in part because provosts are going to insist on solving it. As the stakes get higher, they will figure out how to get together.

The basic rules haven't changed at all. Give the faculty what they want, and the institution will support the library, more or less. Fail to do so, and it won't. The faculty mostly don't care about whether the library owns the information, as long as it's easy for faculty to find the information. It is my sense that most academic libraries have already figured all of this out, at least implicitly. They want to be the information retrievers in the information age. It seems to me that that is just the right line of work to be in.

NOTE

1. John Seely Brown and Paul Duguid, "Universities in the Digital Age," http://www.parc.xerox.com/ops/members/brown/papers/university.html

NORTH AMERICAN LIBRARIANSHIP: A COMPETITIVE ADVANTAGE

Robert Wedgeworth

When I first met Dick Dougherty, he was already an author, journal editor and a nationally known academic librarian. My first comment to him was that I thought he would be older. I am still impressed by the breadth and scope of his many accomplishments. Over the years our professional and personal lives have had numerous intersections. I will only mention a couple. We first collaborated as journal editors. In the early 1970s, editors of the journals of the divisions of the American Library Association (ALA) had to negotiate all of the arrangements for the printing and distribution of their journals. Dick and I, working with Eileen Mahoney persuaded the ALA to establish a centralized editorial production unit of ALA Publishing to serve the division journals and newsletters that is still operating. We are both very proud of that achievement. It led us both to become more involved in publishing activities than we would have ever imagined. Also in the mid-1970s, at Dick's invitation, we began to share a family vacation in Aspen, Colorado. Although we vacationed separately after my daughter started school, our families continue the tradition.

For the generation of librarians who began their careers in the late 1950s and early 1960s, it would be difficult to find a greater role model than Richard M. Dougherty in terms of leadership, change, initiative and risk-taking. Therefore, it is an honor for me to participate in this symposium honoring him for his many contributions to academic librarianship, higher education and publishing. The symposium organizers asked me to offer a perspective from the profession. In doing so, I would like to explore a theme with you that I have begun to develop. The question is "why is North American librarianship

Advances in Library Administration and Organization, Volume 18, pages 287–291.
2001 by Elsevier Science Ltd.
ISBN: 0-7623-0718-8

different?" And to the extent that it is different, "how does that become a competitive advantage?"

There is no doubt that North America – the U.S. and Canada – is the largest and strongest regional library community in the world. In three systematic surveys of the history, status and condition of libraries and librarianship that I have conducted since 1980, there is no region that comes close to matching the resources and capabilities of North American libraries.[1]

There are many possible explanations for this. Perhaps, the most fundamental factor is that libraries have developed in this region as one of many responses to the needs of a democratic society.[2] When books and libraries were scarce and expensive, libraries were a collective response to a variety of educational and recreational needs of growing democracies. However, as books and journals became more affordable, the investment in libraries continued to grow at even a greater rate after World War II.

Another factor for consideration could be the absence of war. Not since the Civil War, over 130 years ago, has there been any significant destruction of public facilities due to war in North America. This period of uninterrupted growth and development cannot be matched in Europe, for example, where two major wars and other smaller conflicts have destroyed many libraries and hampered growth, most recently in the former Yugoslavia. In addition to libraries destroyed, many collections were dispersed for safekeeping in regions like the Soviet Union, never to be returned to their original homes. Others were "liberated" by the victors like the rare book treasures of Germany that are now claimed by Russia.

The relative wealth of North America is a more obvious explanation for its well-developed library community. Certainly, early growth of libraries was stimulated by the philanthropies of Andrew Carnegie who challenged communities by building academic and public library facilities that they were required to stock and maintain.[3] But wealth is unevenly distributed across our region, and in even the poorest regions there are libraries comparable to what is available in many other nations.

It might be expected that the well-educated population of North America might have demanded superior library service to meet user needs. Yet there is little evidence of a popular demand for library service except at basic levels. Indeed, it was well past the half-century mark before there was any major national policy for financing libraries in either the U.S. or Canada.

An advanced international publishing and communications industry would also appear to be a healthy stimulus for library development. Until recently, the publishing industry in North America issued more new publications than the aggregate of the European Community. Except for efforts to establish federal funding to purchase books, publishers have been relatively weak advocates for libraries. The communications industry is divided in its support for libraries. Some communications firms have allied themselves with libraries to oppose intellectual property legislation or censorship issues. Others continue to view libraries as entities that are likely to be replaced by new technologies. Three distinct characteristics appear to set North American libraries apart from their counterparts: Professional education, library cooperation, and the specialized industries that have grown with and supported libraries.

Since the 1920s the U.S. and Canada have maintained a standard for professional that requires librarians to be college graduates and to have completed a post-graduate degree course in librarianship based in a university. Although there are no government certification requirements for individuals other than school librarians, the majority of working librarians in North America meet this standard. Despite the concerns in the 1980s about the demise of library education, the programs have continued to evolve and are stronger today than at any other point in their history. An important change in the professional education of librarians and information specialists has been the shift in the leadership programs from private universities like Chicago, Columbia and Emory to public universities like University of Illinois, University of North Carolina and the University of Michigan. This occurred at a time when a number of small academic and professional programs made this shift due to a widespread effort to down-size private universities programs.

Librarians in many other parts of the world have varying standards for professional education or none at all. Most do not require a tertiary degree from a university and most of the professional training courses for librarianship and many other fields are based in technical institutes rather than universities. While this is changing in many countries, the pace is very slow. In the United Kingdom, for example, it was not until the late 1970s that a university postgraduate degree course was required for librarians. Recently, Norway, Sweden and Denmark have incorporated library training programs into their universities. This convergence of programs for educating library and information professionals in the U.K. and northern Europe should have an enormous impact on the quality of future professionals in these countries.

The second distinguishing characteristic is a long tradition of library cooperation. From the beginnings of the organized profession in North America cooperative activities have characterized the work. The first interlibrary loan code dates from 1917.[4] Originally, based on the work of professional associations, these activities have grown from union lists, resource sharing and joint reading guidance lists to major library utilities, shared operating systems and joint facilities. While numerous countries publish national bibliographies that support local libraries, none have achieved the operating economies and dramatic expansions of service that have resulted from OCLC, RLG, WLN and numerous state and provincial library systems.

Library cooperative activity assumes that its participants share a common understanding of the field. Lack of trust and lack of a common base of professional education appears to hinder even the most fundamental cooperative activities in other countries.

A third distinguishing characteristic of North American librarianship is the specialized industry that has grown concurrently. The early creation of specialized library book publishing (Library Journal) and distribution, library suppliers and specialized library furniture have been significant contributions to the growth and development of the field. This market that is now international, produces major economies of scale. Intelligence carried from one library to another by salespersons supplements the communications that occur at library conferences where products and services are exhibited. The large geographical distances in North America and its transportation system contribute to the development of this market. Through the leadership of North Americans, the market for information products and services sold to libraries has expanded dramatically. Early 20th century reports of librarians attending professional conferences are full of comments attesting to the importance of being able to consider so many library products and services by attending a national conference.

Whatever the extent to which we can establish these three characteristics of North American librarianship as significant, they become competitive advantages as we look toward the future. As all library communities strive to maximize their resources and capabilities for the populations they serve, those that adapt themselves through innovation and cooperation are likely to be the most successful. Currently, the European Community is developing a common library development program that will rival developments in North America if language and cultural differences can be bridged. Although government financing drives this development and pressures to serve more users encourage libraries, the cultural and educational differences among European librarians will continue to hinder progress. What seems likely for all libraries is that the cost of "contents" regardless of format will continue to escalate at substantial rates. Also, technical expertise required to maintain library systems would be increasingly scarce and costly. Those library communities that are adept at fashioning collective responses to these pressures are most likely to be successful in the immediate future. Using the leverage of multiple institutions to gain price and service adjustments will become necessary for libraries to do more than just survive.

The North American library community that produced Richard M. Dougherty, researcher, educator, library association leader, publisher and advocate for scientific management has also produced a cadre of similarly educated professionals capable of responding to his leadership and initiatives. It is a singularly unique contribution to our field worldwide.

NOTES

1. Robert Wedgeworth, ed., World Encyclopedia of Library and Information Services, 3d ed. (Chicago: American Library Association, 1993).

2. Sidney Ditzion, Arsenals of a Democratic Society: A Social History of the American Public Library Movement in New England and the Middle States from 1850 to 1900 (Chicago: American Library Association, 1947), pp. 51–76.

3. George S. Bobinski, Carnegie Libraries (Chicago: American Library Association, 1969).

4. World Encyclopedia (1st ed.), p. 6.

SOME REFLECTIONS ON UNIVERSITIES, LIBRARIES AND LEADERSHIP

Billy E. Frye

A version of this paper was presented as the openning address of the Frye Leadership Institute of the Council on Library and Information Resources, held June 4–16, 2000, at Emory University, Atlanta, Georgia.

INTRODUCTION

I have been asked to address in this paper issues relating to universities, libraries, and leadership. I claim no expertise about any of these things. But after almost 50 years of association with them as student, faculty member and administrator, I do confess to a special concern for them. As for my own role, having very little real idea of what challenges the future holds for you who are rising leaders in information management, I can only hope that you will extend to me the same charity that was extended to the young Quaker farmer at a community gathering. He had some reputation as a vocalist, and was asked to sing after dinner. As it happened, he was as hoarse as a frog, but he agreed to try. He struggled along for awhile, but finally broke down altogether. "Never thee mind lad," said a consoling elder. "Never mind, thee's done thy best. But the fellow that asked thee t' sing ought t' be shot!"

Advances in Library Administration and Organization, Volume 18, pages 293–305.
Copyright © 2001 by Elsevier Science Ltd.
All rights of reproduction in any form reserved.
ISBN: 0-7623-0718-8

UNIVERSITIES

I am going to begin with some comments about colleges and universities because they, particularly the so-called research universities, are what I know best. But I also place my remarks in this context because if universities are not in actual fact the cutting edge of experimentation and reform in scholarly information management, they ought to be – indeed, they must be. There is no other context within which we can hope to sustain a coherent approach to the processes by which knowledge is synthesized, conserved and disseminated with some semblance of order, comprehensiveness and integrity, relatively free of the distorting influence of special interests.

Colleges and universities are wonderful places. They sustain a culture of intellectual freedom, skepticism and inquiry that is essential not only to the intellectual life, but also to our democratic and economic ideals. Who can question that our success as a nation is closely tied to the fact that we have created in this country the most successful system of higher education in the world.

Successful, yes, but not perfect. The mission and purposes of colleges and especially research universities are generally very complex. The variety of programs and constituencies that they support are very great compared to other institutions. In combination with the large size of most higher education institutions today, this complexity poses enormous challenges in governance, goal and priority setting, communication, evaluation, and in generating a truly integrated community. This is probably the most significant factor behind the well-taken lament of Brian Hawkins and Patricia Battin (1998) when in *The Mirage of Continuity* they asked, "why is the transformation process in higher education so slow, so disorderly, so expensive, and so resisted?" (p. 4–5).

A partial answer, at least, has to do with the excessively fragmented, compartmentalized and autonomous way we have organized ourselves to deal with scholarly specialization and institutional size. Undoubtedly this is also the source of our great strength, but as so often seems to be the case, the source of strength seems also to have become a serious liability. As in so many other things, obsession with turf, and all that goes with that – professional identity, rules and standards, prestige, access to resources, etc. – is a nearly insurmountable problem when we contemplate the need for quick, coherent responses to the challenges of our time.

In any case, though remarkably effective centers of social change, universities themselves tend to undergo overt change, as Jonathan Cole (1993) has noted, only when there are large and persistent disequilibriums in economic or social conditions. I would add to that the changes that come when a significant epistemological or methodological breakthrough occurs, as with some scientific discoveries and technological developments. This is such a time. The challenges that have arisen in recent years are by now very familiar to you. Over a decade ago the then sluggish and unpredictable economy brought an end to the era of uninterrupted growth that we had enjoyed for the prior 3 or 4 decades. The political climate since then has made it clear that research universities will have to exploit new sources of income to sustain their mission and ambitions. Intractable social problems like violence, drugs, ignorance, and poverty have shifted public priorities away from higher education to more urgent matters. Yet, at the same time the public began to hold universities more accountable, both for their own conduct, and for solutions to those problems.

Added to these pressures for change are the challenges of globalization, and even the evolution of the academic disciplines themselves as knowledge has expanded and become interconnected at an unprecedented rate. And of course information technology is changing dramatically how we handle information and how we generate, evaluate, organize, preserve and disseminate scholarly knowledge in our teaching and research. Perhaps most significant of all, information technology is quickly rendering obsolete the traditional disciplinary boundaries around which the university is organized, and through which we organize and access knowledge. Though not yet widely recognized in practice, it has already rendered inadequate our current budgeting practices and tenure and other personnel processes.

Because of these pressures, a lot of changes are already occurring. I won't try to recount these, but two things can be said about where things now stand. First, no one as yet can say with any certainty what the "new paradigm" of the university will be. According to many observers, the successful university will likely:

- learn to live with low or no-growth budgets, substitute new sources of income for old ones, and increase productivity without the benefit of governmental beneficence;
- give more attention to teaching undergraduates, and provide stronger incentives and better rewards for good teaching and for those who give more time and attention to students;
- develop a new understanding of teaching and learning, and of the relationship between teacher, learner and institution;
- be more selective and focused in the range of its academic programs, and find ways to compete in the academic marketplace through market differentiation and niche specialization;
- reduce the hegemony of the traditional departments and schools in favor of a greater degree of cross-disciplinary, integrative scholarship and greater prominence of collegiate objectives;

- use information technology not just to increase the effectiveness and efficiency with which we do old things, but to find radically *new* ways of creating, organizing, authenticating, accessing, and disseminating knowledge and of credentialing those who discover and teach.

But however valid these characterizations may prove to be, they fall short of a coherent picture of the future because they don't get to the heart of what matters. That is how a purposeful intellectual community will be nurtured in the digital environment. That this is NOT an easy problem is suggested by Brown and Duguid, in their essay, "Universities in the Digital Age" (1996), where they write, "The notion of a virtual campus both underestimates how universities as institutions work and overestimates what communications technologies do. . . . Learning involves inhabiting the streets of a community's culture . . . (and) . . . experiencing its cultural peculiarities" (p. 13). We must always keep this in mind as we fantasize about the grand sweep of the digital era.

So, the second thing one can say about the current state of change in colleges and universities is that up until now, at least, we seem not to be fully up to the task of adapting to the new technological world we have entered. This is not surprising, of course. By and large our response to our changed environment seems to be directed more at survival and maintaining the status quo – doing old tasks in new ways – than at deep reconceptualization. Few if any major institutions have begun in more than a tentative, piecemeal way to explore the challenges and opportunities of the "new university paradigm" that Jim Duderstadt (2000) and others have talked about so effectively.

The principal point I want to leave you with is this: it is inevitable that in today's world our leaders will continue to be preoccupied with such things as competition and prestige in the "academic marketplace". The context of this competition will continue for a time to be shifted to issues like intellectual property, distance learning, electronic publication and other challenges and opportunities arising from digital technology, as well as other familiar themes such as technology transfer and alliances with industry. But as important and seductive as these things are, they should be understood as means, not ends. And so we should be clear about what it is we want to accomplish.

Perhaps I can make my meaning clear by example: a few years ago I undertook to talk to several hundred Emory faculty, staff, students and alumni about what they thought would be most important to Emory as we entered the next phase of our development. Overwhelmingly, their principal concerns had to do with such matters as these: preservation of our commitment to teaching in a climate that increasingly rewards research; promoting interdisciplinary collaboration; preservation and enhancement of a sense of community, and the reflection of this in the physical structure of the campus; increasing the reach and effectiveness of our collaborations with the communities around us; and, of course, maintaining the highest standards of intellectual excellence. Implicit in this response was an understanding of the key problem in planning, and that is that without vigilant attention to underlying institutional values, the pressures stemming from the exigencies and opportunities of these times will divert us from our fundamental mission, or distort it beyond recognition. The greatest responsibility of leadership is to create a solid, consistent linkage between the changes that are needed and the mission and basic values of the institution.

LIBRARIES

But before I come to that let me turn to a few reflections on libraries and information management. Victor Hugo observed that "all libraries are acts of faith" (Fusi, 1989, p. 6). I like this because it evokes not only the idea that the future is intimately connected to the past, but that a visionary commitment is necessary to work through the current challenges faced by libraries and their partners as they continue to write the story of mankind. (At this juncture I should parenthetically note that I am very prone to use the word, "library", when I mean information resources in general. I assure those of you who are not librarians that I recognize and fully appreciate the essential and increasing role that information technology plays in virtually every aspect of information management.

Thus qualified, several familiar images suggest the central role of the library: the 'gateway to knowledge'; the 'epistemological center' of the university; the 'cathedral of learning'. A biologist like myself might see the library as the 'organizing principle' that brings order to the forming embryo of knowledge. But perhaps the best metaphor is that of the library as 'narrative' – the narrative of cumulative scholarship. A people or a culture exists over the long haul only in the narrative that it creates about itself. Whether myth or real history, fact or fiction, this narrative is embodied in its totality only in the library. How well the pieces of that narrative lend themselves to being understood as an integral story; how effective it is in shaping and reshaping our self-understanding depends upon the twin ideas of the library: *order* and *accessibility*. Notwithstanding the current ambiguity about the physical and administrative boundaries of the library, these remain the basic concept of the library, now and in the future.

I used the word ambiguity just now because there *is* currently some unfortunate ambiguity or ambivalence about the role of the library that needs to be recognized. I believe this can be attributed to at least three things. The first, of course, is the loss of discrete boundaries with the advent of digital technology and therefore of a clear physical delineation of the library. This goes hand in hand with the proliferation of a confounding array of information sources, and the library's loss of control over much of it.

Second, with this proliferation and our almost drunken infatuation with the growing plethora of "information" has come some conflation of means and ends. The administrative organization that defines the library's relationship to other information resources and user communities is, or should be, only a means to the end, yet we are continuing to let traditional structure rather than functional objective shape our objectives and institutional priorities in the arena of information management.

Third, with today's predominant emphasis upon new facts and data and the use of information for immediate personal and social ends, we seem to have de-emphasized the importance of the long term processes of critique and synthesis by which the corpus of knowledge is molded into ideas of lasting significance. Thus we have created an ethos in which the immediacy of information has taken precedence over long term, organized knowledge which is the forte of the library. It would be too much to claim that libraries are, to use E. O. Wilson's (1998) recent term, "consilient" repositories of knowledge, though they are latently so. But it *is* defensible to argue that it is only through partnership between the library and the inquiring, skeptical scholar that actual realization of the vision of consilience of knowledge is imaginable.

When I spoke at the Harvard "Gateways to Knowledge " conference several years ago, I said that the principal challenge could be summed up in two points: First, it is untenable for college and university libraries to meet future information needs of faculty and students solely through the traditional avenue of growing their collections. Second, new digital and telecommunication technologies offer the possibility of resource sharing and collaborative collection development and management that were unimaginable only a few years ago. Thus, cooperation is the only realistic way for universities to have access to a comprehensive source of knowledge in the future as in the past. This is still the central challenge, I think, and is embodied in the vision of "the virtual library" or "the digital library": the dream that through the powers of computer and telecommunication technologies, the libraries of the nation and the world will eventually be linked to one another in a transparent network that will enable users to have access to any information in any format anywhere, anytime, instantly and at an affordable cost. Yet today, notwithstanding the great progress that has been made in things digital, we seem not to have come much closer to the real change of paradigm envisioned in the notion of the virtual library than we were a decade ago. So we must ask "why not?"

It is *not* for lack of technical capacity. I am sure there are some technology-related issues, among them the very modest amount of the scholarly record that is in digital format in many fields; problems of copyright and ownership, related legal and economic issues; and, the continuing instability of the technology environment that makes us uneasy about committing to something that is not yet predictable or controllable. But I would suggest that the failure of the pieces of the "digital library" to fall into place as yet is due mainly to three interrelated problems:

- the powerful grip of the traditional model of institutional autonomy that maximizes individualism and almost precludes collaboration, if it entails a sacrifice of institutional autonomy;
- therefore, in well intended consortial efforts, we have repeatedly reduced the arena of collaboration to the lowest common denominator of shared interest among the participants; and,
- the diversion of collaborative efforts to the working out of local problems, technical, economic or otherwise, rather than to the creation of a new approach to information management.

Let me turn now to the subject of leadership, which may be the most important and difficult challenge of all.

LEADERSHIP

There is a story that when Otto von Bismarck took command of the Prussian army in 1871 he was looking for leaders. So he reviewed the personnel files of all of the officers and divided them into four categories. In the first category he put those who were capable and ambitious; in the second, those who were capable but not ambitious; in the third, those who were neither ambitious nor capable; and in the fourth, those who were ambitious but lacking in ability. The first group he promoted to positions of high responsibility as quickly as possible. The second he put into middle management where responsibility but not initiative was needed. The third he put in charge of remote outposts of the empire that really didn't matter much to anybody. And the fourth group he took out and shot!

Recently I was asked by the President of a distinguished university what I thought the essential qualification of the next librarian was, and I responded as follows: "It is a given that the Librarian must have strong managerial, budgetary, organizational and planning skills. He or she should be well versed in all the principal issues and trends of the day, and well connected to the professional community. He/she must have the knowledge, judgment, interpersonal skills, and personality to command respect and support from both the local university community and the larger library community; must understand the nature of great research universities and have the political sensibilities to work effectively within it and garner the confidence and loyalty of staff, faculty and administration; must be frank and open, yet patient and diplomatic; fair and honest; instinctively consultative, yet decisive; a team builder, communicator and teacher. He/she must be both pragmatic and visionary, capable of articulating a compelling view of the library's essential place in the future of the university, and of conceiving the practical steps to that role."

Spun out this way, this characterization of the demands of leadership sounds ludicrous and impossible to attain, though I do think it fairly characterizes our expectations. So I would like to make just a few points about the challenges of leadership that may be worth reflecting upon as we contemplate the perpetual problem of leadership in our tradition-bound but changing institutions:

First of all, the need is great and the demands are extraordinary. It has been said that the two essential ingredients of effective leadership are: (1) the ability to craft a coherent vision, and (2) the ability to persuade talented people to embrace and pursue that vision. In the college and university context, think at least three other ingredients have to be added: (3) the ability to empower people to act despite sometimes overwhelming institutional resistance; (4) the capacity to deal with ambiguity and uncertainty in the face of a growing demand for accountability; and (5) the ability, perhaps instinctive, to avert what I will call the "power paradox", that is, to provide the essential foundation for confident action without becoming identified too much with trappings of power and self-interest.

My second point about leadership is related to the first, and it is that different situations require different kinds of leadership. This is obvious, but a view that puts it in large perspective can be adduced from Pierre Levy's discussion of "anthropological space" in his new book, *Collective Intelligence: Mankinds' Emerging World in Cyberspace* (1997). Levy suggests four images of the 'spaces' we occupy and by which we identify our place in the world. The first he calls earth space. This is the space in which we position ourselves in relationship to our ancestral line, to the earth and the cosmos. Totemism is an example of earth space. The second is territorial space, the states, institutions, governments or other political systems with which we are affiliated and from which we derive a part of our identity. The third is commodity space, which can be thought of as the profession, function or activity by which we identify ourselves – it's what we do. And finally, Levy suggests something is now emerging that he calls knowledge space. This space defines our membership in the growing community of knowledge, the 'collective intelligence' of mankind that Levy sees emerging in the digital environment. As human society has evolved, each successive space has not necessarily replaced the earlier one, but has been layered over it. All of them can be found in our sense of our identity, which is dominant depending upon the nature of our particular society.

Leaders emerge in each of these anthropological spaces. One might think of the tribal shaman or religious priesthoods in earth space; the military statesman in territorial space; and the business tycoon (and, if I may be so cynical, the politicians they buy!) in commodity space. Today and historically we have had a mix of all these, both as embodied in our individual leaders, and in the aggregate profile of our leaders across the spectrum of society.

The question, of course, is what kind of leadership will emerge in knowledge space? Levy points out that our traditional decision-making processes evolved in a relatively stable social environment, one that responded well to the predominantly unidirectional flow of authority. Digital technology has greatly destabilized that environment. It has done so by increasing the amount of information available to decision makers; by creating multidirectional communication; by increasing the level of democratization and the potential of anonymity.

In this environment will a knowledge-based meritocracy really emerge, as some suppose? Will control be assumed by those who have best access to and command of knowledge and ideas, or only to those who have the greatest capacity to manipulate and control them? Or is it possible that something entirely different will emerge, a leaderless society in which decisions are truly made and directions set by the force of reasoned knowledge and networked consensus? Levy doesn't explicitly say so, but he seems strongly to imply movement in that direction. However, while I believe and hope that the emphasis upon meritocracy and upon networked, consensual decision making will increase, I do not believe that our dependence upon personified leadership will decline. I have two reasons for this opinion: the first has to do with the basic biological nature of leadership. In her book, *Danger in the Comfort Zone* (1995), Judy Bardwick points out that when all is said and done, leadership is not intellectual or cognitive; it is emotional. People accept and respond to leaders because they have an emotional need for leadership, for someone whom they can trust and depend upon and to whom they can make a commitment. In short, we intrinsically need leaders. The second has to do with a distinction between the personal engagement, the intelligence and creativity that is necessary to generate interesting ideas, and the capacity to motivate and organize people, and to cause a sense of order and purpose to coalesce around an idea, that characterizes leadership.

My third point about leadership is that the problems that leaders in the arena of information management face are huge. The biggest problems concern not so much knowing what to do, as coping with the environment in which it must be done. It has been my privilege over the past decade or two to talk with

many highly qualified, dedicated librarians, and in several instances to participate in the evaluation of their institutions. Almost all were attempting to cope
with a constellation of issues, something like the following:

- *budgets that were inadequate in size*, and structured in a way that did not
  permit the librarian and her/his staff to make the choices that they deemed
  to be in the long-run best interests of the institution, or to manage cost-
  effectively;
- *lack of a clear sense of priorities* or direction from the central administration of the university by which to guide their own decisions. This is a situation
  which effectively leaves the library trapped between the highly compartmentalized vested interests of the university, and the demands and
  opportunities for cooperation that are being created by the networked, digital
  environment.
- *A clientele* that is extremely diverse in its needs and capabilities, and sometimes very demanding and inflexible. In part because they, too, do not see
  the larger institutional picture because of the fragmented structure of the
  university;
- *A culture* that still seems to value and reward the wrong things. Only a few
  years ago I visited a major university where one of the stated objectives was
  to develop collections of comparable size and comprehensiveness to those of
  peer institutions with much larger and older libraries. As anachronistic as
  such a goal now seems, it obviously reflected the belief that acquisitions and
  collection size, not access to information, would continue to be the standard
  by which both individual performance and institutional reputation would be
  judged.
- *Lack of effective coupling* with, if not outright isolation from, the top level
  decision makers. This results in ambiguity about the real level of responsibility and authority vested in the librarian and the question whether the library
  is central to the plans of the university. Such a solid thinker as Jerry Pelikan
  (1992) – not one we would expect to be swept away by the fascinations of
  bureaucracy – says that "there is a need for the university library to be
  involved as a genuine and full partner . . . in both short-range decisions and
  long-range planning" (p. 117).

CONCLUSION

Well, I must bring this to an end, but truthfully I find it difficult to draw practical conclusions, so I am tempted to fall back on an allegorical story that I
once heard Vartan Gregorian use in a similar situation.

Once upon a time, there was a bunny rabbit who became very tired of his role as one of Mother Nature's shyest, most timid and vulnerable creatures. So he decided he would change. But what should he become? He thought and thought, and then in an uncommon surge of inspiration he hit upon an idea: "I will become a raccoon," he thought. "The raccoon is a bold, intelligent, inquisitive creature. He does whatever he wants. And besides he has a splendid coat that would be just perfect for those late fall football games!" The more he thought about it the more excited he became. But being prone to misgivings as he was, he began to think, "This is a very big step in my life. Maybe I should consult my friend, Wise Old Owl." So he went to Owl, told him his plan, then waited anxiously for Owl's approval. Owl listened attentively and then after several minutes of silent reflection, he said: "Well, this is a worthy plan, indeed. But as long as you are changing, why don't you set a higher goal for yourself. The raccoon is nice, but he isn't the strongest, noblest creature in the forest. So long as you are changing, why don't you become . . . well, why not become an eagle?" This bold, visionary suggestion hit bunny like a bombshell, and he became so excited that he could hardly contain himself. Jumping up and down, he exclaimed, "Yes, yes!! I'll become an eagle! But tell me, Owl, how do I do it?" "How should I know?" replied Owl. "I only give policy advice!"

Well, neither do I have any prescriptive proposals to effectuate the changes to which I have alluded today. But I would like to end by underscoring several changes that would lead, I believe, in the right direction. Let me put it this way: When I wake up tomorrow morning and look into the mirror of academe, what would I most like to see?

First, I would like to see colleges and universities engaged in a form of planning that decidedly places primary emphasis upon our values and our vision of what they ought to be, and secondary emphasis upon so-called strategic planning aimed at the exigencies of the environment and imagined scenarios of the future. Only then, I think, will we recognize that Newton's 3rd law applies to the life of institutions – whether intended or not, every proximate action that we take has some ultimate effect upon the direction in which we move, and all decisions should be carefully considered in that light.

Second, I would like to see provosts and presidents taking steps to relax the hegemony that our compartmentalized structure gives to departments, schools, and other vested interests. I do not suggest that they should be eliminated or even weakened; but I do believe that our image of the university should be one of an intellectual tapestry made stronger and more interesting and versatile by the interweaving of the woof of interdisciplinary dialogue and collaboration across the strong disciplinary warp.

Third, I would like to see libraries and other "information managers" positioned at a pivotal and influential intersection in this network and given the voice and attention that behooves the curator of our most important resource – knowledge.

Fourth, I would like to see a coalition of major universities, library and information resource organizations, and foundations recognize the urgency of the leadership problem. I would like to see them establishing concerted programs to recruit more of our brightest colleagues, especially those with the necessary vision and skills, into careers in information management. And I would like to see them building the interdisciplinary teams that everyone agrees are necessary to deal with the challenges of the digital era.

Fifth, I urgently want to see a small and determined cohort of our leading universities make a renewed commitment to actually bring the "virtual library," or a compelling prototype of it, into being! I emphasize a coalition of universities, and not just libraries, because I think success depends upon an institutional commitment – upon the determined commitment and persistent attention not only of librarians, but also of presidents and provosts. Then and only then will the stumbling blocks of institutional structure, autonomy and tradition be removed and the elements of the virtual library (that are mostly already there) fall into place as a functional entity. Only then will universities, in the words of Richard Katz (1999), "develop the consortial relationships and create the organizational incentives and structures . . . [that are needed] . . . among librarians, technologists, media specialists, faculty members and others."

To what extent this kind of change will require formal structural and policy changes, I do not know, though surely strategic adjustments in policy and organization must accompany the transformation that we seek. But I am convinced that one crucial need is for leaders who can help us develop and communicate the vision, conceive the strategies, change attitudes, motivate our communities, and eventually even modify our culture in order to bring about the changes that are needed. I have great hopes that those of you assembled here will make important contributions to meeting this need.

With that, I will subside and hop back over the fence to see what's growing in Mr. McGregor's garden. Thank you for being here, and for letting me participate in the inauguration of this new experiment in leadership in this way.

REFERENCES

Bardwick, J. (1995). *Danger in the Comfort Zone: From Boardroom to Mailroom, How to Break the Entitlement Habit That is Killing American Business*. New York: AMACOM.

Brown, J. S., & Duguid, P. (1996). Universities in a Digital Age. *Change, 28*(4), 10–19.

Cole, J. (1993). Balancing Acts: Dilemmas of Choice Facing Research Universities. *Daedulus*, *122*(4, Fall 1993), 3–36.

Duderstadt, J. J. (2000). A Choice of Transformations for the 21st Century Library. *Chronicle of Higher Education*, *46*(22) (February 4), B6–7.

Hawkins, B., & Battin, P. (1998). *The Mirage of Continuity: Reconfiguring Academic Resources for the 21st Century*. Washington, DC: Council on Library Resources and Association of American Universities.

Katz, R. N. (1999). *Dancing with the Devil: Information Technology and the New Competition in Higher Education*. San Francisco: Jossey-Bass.

Levy, P. (1997). *Collective Intelligence: Mankind's Emerging World in Cyberspace*. New York: Plenum.

Pelikan, J. J. (1992). *The Idea of the University: a Reexamination*. New Haven: Yale University Press.

Wilson, E. O. (1998). *Consilience: the Unity of Knowledge*. New York: Knopf.

THE RESEARCH LIBRARY DIRECTOR: FROM KEEPER TO AGENT-PROVOCATEUR

Paul H. Mosher

President Charles William Eliot of Harvard University was crossing Harvard Yard in 1858 and encountered librarian John Langdon Sibley, who had a smile on his face. Eliot greeted him and asked where he was going. Sibley responded, "Everything is in order Mr. President. The library is locked up and all the books are in it but two, and I'm on my way to get them now".[1]

This wonderful anecdote epitomizes what I will call the "first stage" of the evolution of the concept of the university librarian during the past century and a half, a period during which the interactive worlds of publishing, scholarly communication – and of academic libraries – was fundamentally transformed. There appear to be four stages in this evolution, which may give some coherence and shape to the transformation of the university library director's role as an actor and reactor to change in higher education and publishing over the last century or so. These stages may be periodized as follows:

I.   To 1928: The "keeper", evolving to the "professor-librarian".
II. 1928–1970: the collector or "bookman-librarian".
III. 1970–1985: the organization man (or woman) or "scientific" librarian.
IV. 1985–present: the provocative or "networked" librarian: the "change-agent".

These distinctions should not be taken as exclusive for, in real life, distinctions blend, move, disappear and reappear. But this rough chronology may prove

Advances in Library Administration and Organization, Volume 18, pages 307–316.
Copyright © 2001 by Elsevier Science Ltd.
All rights of reproduction in any form reserved.
ISBN: 0-7623-0718-8

useful in giving some shape to the development of the library director's self-concept and focus over the last century-and-a-half.

I. From ca. 1850 to 1928: The "Keeper", or "Professor-Librarian".

The concept of the library director as a "keeper", in the sense that Harvard librarian John Langton Sibley would have understood it, goes back a long way. In 1650, John Dury (a Scottish clergyman and educational reformer) published *The Reformed Librarie-Keeper* in 1650 (London). In it, he opposed the current concept of the library as a cabinet of curiosities, including books, and he criticized the library of the University of Heidelberg, writing that "they that had the keeping of this librarie made it an idol, to bee respected and worshipped for a raritie by an implicite faith".

Many academic libraries on both sides of the Atlantic carried out the functions of natural history museums (cabinets of curiosities) in which books were, presumably, also considered "curiosities" as well as objects of study, and thus the heads of these institutions were considered "keepers", to protect books as well as museum objects. As late as 1891 the new library on Penn's new campus in West Philadelphia was designed as a late-blooming example of the hybrid of library and archaeological museum.

The librarian of those days was usually a faculty member, an amateur librarian serving part-time, and his role was to keep and protect the books (which came to the library by gift and faculty purchase) and organize them for use by faculty, not students. Most university libraries of the time – and there were many fewer than today – resembled small college libraries in scale, and had only a handful of staff. Of course, the universities themselves were mostly tiny by later standards. The Librarian, acting as a head clerk, was always male; the clerical assistants, often female. A typical conceptual view of the academic library of the day was Otis Robinson's *How a College Library Shall be Prepared for Use.*[2]

As the first library buildings developed, at Dartmouth or Cornell, for example, they reflected the static and protective permanence of the Librarian's concept of his role. Each library function had a fixed and permanent room – the function and scale of which were expected to remain constant over time; there was no anticipation of the need for future change or alteration in functions, or in the spaces designed to hold them. Permanence, stability, dependability, security, control, were the prevailing values.

The escape of the university library from the "cabinet of curiosities"[3] – separation from the museum and the building of a set of new, larger, separate library buildings, corresponded to the first wave of development of the graduate and professional "research" university, and a new concept of the position of university librarian developed.

II. 1928–1975 the age of the "Bookman-Librarian"

The director in this long period was more often a full-time director, though often still recruited from the ranks of the faculty. There was, however, an increasing number of professionally educated librarians – especially after World War II. The library director's role was understood primarily as that of a "bookman" that of a collector-writ-large (often personal as well as official), who belonged to book clubs and courted great and wealthy collectors in hopes – often successful – of snaring their collections or collection endowments. He – it was still "he" until World War II – was often a stern, paternal figure, who believed in tight oversight of decisions; and who paid a lot of attention to process, and functional and budgetary control, in detail. He tended to be hierarchical and judgmental in character and values, and was a grower of collections, an expander of services, and a grower of budgets. "More" was a valued keyword.

Guy Lyle identified the major shift from the Professor-Keeper Librarian "in the later 1920s, when the boom in college enrollments and curricula precipitated a revolution in library use and a need for more systematic and orderly methods of organization and reader assistance". He then went on to describe what he then saw as the "modern" librarian as "the new professional bookman-librarian".[4]

To be sure, there were wonderful figures like Lawrence Clark Powell at UCLA, and Bob Vosper at Kansas, who were hardly stern, and who were beginning to rethink the traditional roles of university libraries and librarians, and to push for a global dimension to librarianship and library cooperation. The exceptions, as usual, may have proved the rule.

During the Age of the Bookman-Librarian, the focus was on structure, organization, and arrangement – on inputs, accumulation, growth in scale, budget and unit processing efficiency. Services were desk-bound, paper-based, file-oriented, fixed-place and mediated, with the librarian-expert standing or sitting – firmly – between the patron and the information. The finding structures were paper in format, rule-and convention-based, and became traditional and deeply held as values by librarians. Change occurred, but it moved very slowly. The library was not expected to respond quickly to changes in scholarly terminology, research, or new thinking. The structure and rules tended to be authority-based rather than user-based.

"Organization for work" was a popular phrase during this period, and management values were drawn from industry – particularly from Frederick Taylor, the father of "scientific" industrial management – and organizational models tended to be drawn from the military – the organization chart was derived from the military table of organization and equipment.[5] The influence in the U.S. of the industrial revolution and of the world wars played its part on library organization and operation. Productivity goals were based on quantified study of work output, such as the time-motion study to eliminate wasteful effort. The objective, as Dougherty and Heinritz have pointed out, was "to provide service as economically as possible at the level of quality required".[6] Taylor considered the detailed study of processes and the structure of work rather than people, relationships or productive work cultures.

This was the period when Shiyali Ramamrita Ranganathan produced his classic *Library Administration*,[7] in which he set forth a highly crystalline and taxonomic description of the functions of librarianship. He treated the administration of a library as if it were a taxonomy of butterflies or conifers as a highly detailed, meticulously structured, classified, comprehensive organization and description of tasks. Intelligence, initiative, independence or "attitude" were not valued highly, only high-volume, regimented output.

At the same time, the period of the Bookman Librarian produced highly advanced and sophisticated cataloging systems, and the great union catalog movement – a revolutionary collaborative concept which provided the basis of the interlibrary loan movement – itself a revolution in access by everyone to the nation's knowledge and information repositories. It was also the period of the public library movement, sparked and supported by Andrew Carnegie's well-chosen philanthropy, which promoted widespread social learning and literacy, and public understanding grew, in all parts of the country, of libraries as central forces in progress and learning.

Despite these important advances, as Richard Dougherty wrote in 1993, "Over time, a culture evolved that defined the role of librarians as collectors and preservers of the scholarly record rather than as active participants in the research enterprise. As a resource for many and partner of none, the research library became increasingly isolated in a decentralized, diverse and competitive world".[8]

In retrospect, the selection of the factory as a metaphor for libraries, the use of Taylorian control models and of military organization by library directors, while very "modern", helped to create growing unrest within library organizations, where staff often felt over-regimented and undervalued, and helped set the stage for the managerial revolution that followed in the 1970s. The tendencies toward rigid and controlled library organization and management may have helped to delay earlier adoption of user- and outcomes-based design, and the more creative and innovative applications of information technology to libraries which would have provided more effective and strategic responses to the attack on the significance, utility, independence and function of academic libraries by the computer mainframe czars of the late '70s and '80s.

III. 1970–1985 The "Scientific" Librarian. The organization man (or woman) – the mainframe as metaphor.

About 1970, the focus of library "science" was moving from taxonomy and detailed control to the gathering and analysis of empirical data. University library collections, staffs and budgets continued to grow at a rapid rate, and libraries entered into the shadow of the "serials crisis" that made Robert Maxwell a publishing czar. Universities entered a period in which campuses proliferated rapidly, enrollments, resources, and budgets grew steadily in real terms. It was a process that Robert Zemsky and William Massey have labeled the "academic ratchet": There was simply no end to the productive uses that might be made of *more*. Most library directors understood their roles during this period as getting and keeping a larger piece of the growing action.

Students of library management began at this time to champion management procedures based on the application of "modern scientific principles" and methodology to problems of administration to help libraries with the challenges of growth. This meant "nothing on faith", opposed "how we do it good", and it meant substituting empirical study and experiment and the resulting quantified data for comparative measurements of input such as volumes added. The "heart" of scientific management was the physical aspect of human organization "within the framework of individual and social needs and aspirations". While many more far-sighted directors began to experiment with and employ empirical data in addressing their problems, the methodology and its useful application took a long time to penetrate the library culture as it then existed, and has only more recently been applied routinely and widely, with greater knowledge of tools and techniques by managers. Advances in statistical, analytical and representational software allowed the methods and products of empirical analysis to enter, and influence, the mainstream of library operation and thought. Gradually, the application and use of empirical data passed from being regarded as "theirs" to "ours".

The Scientific Librarians began to move processing and control to mainframe computers-, and to experiment with the possibilities of "online" catalogs to gain processing efficiencies – an enormous process of functional and cultural transformation. Gathering, analyzing and using statistical information and empirical evidence – often a product of automated functions – increasingly became the model, if not always the practice. But despite the introduction at a number of universities of new, quantified, "scientific" techniques of measurement and assessment designed to improve efficiency and output, crises were brewing. Perhaps because of outward signs of prosperity, library directors missed the handwriting on the wall, and began to be threatened from without and within. The "crisis in library leadership" became a common theme in the library literature of the later '70s.

The rapid growth of collections, staffs and budgets had begun to alarm campus managers on the one hand, and the introduction of new and only partly understood computerized operations bewildered, threatened and demoralized library staffs. Failure of "scientific" directors to select carefully, train new staff, retrain existing staff and to involve both old and new staff in the direction and changes being implemented, brought to a head tensions that had been building in those who felt alienated and threatened by decisions and outcomes of which they had often been poorly informed, with which they had been little involved, and which they often did not understand.

A freely adapted section of Kenneth R. Shaffer's "The Library Administrator as Negotiator: Exit the 'Boss' ".[9] may illustrate how the university librarian of the time often felt about his role:

The Tasks Of The Library Administrator:
To decide what is to be done, to tell someone else to do it, to listen to reasons why it should be done by somebody else, or why it should be done in a different way, or why it should not be done at all, and to think up arguments as to why it should be done anyway.

To follow up to see if it *has* been done, to discover that it has *not* been done, to listen to why it has not been done, and to think up arguments why it must be done anyway.

To follow up a second time to see if it has been done, to find that it has been done incorrectly, to point out how it should be done.

To consider how much better it would have been done if he had done it himself, and that he has spent two days discovering why it has taken somebody else three weeks to do it wrong, and to realize that such an idea would be demoralizing to library staff because it would strike at the very foundation of the belief of library staff that a director has nothing to do.

While library directors were thus enmeshed in the complexities and challenges of rapid organizational growth and change, they came under fire from university administrators for allowing costs to become uncontrolled (this was the beginning of the "library as black pit" era), the ARL Library Management and Review Program (MRAP) – highly participative and not closely controlled by directors – began to be applied at a growing number of university libraries in an effort to help staffs understand and address change and growth. The process rapidly brought to the surface of the organization staff strain, resentment, tension and repressed fears that had been growing over the years. Ed Holley, Arthur McAnally and Robert Downs discovered an uncommonly high turnover rate of university library directors that resulted during the "MRAP years" well over 22 directors retired or were moved on – representing over 25% of the libraries which were then ARL members.

At the same time, computer center directors laid claim to the future of information by capturing the term as a synonym for technology, and claiming that the information power of technology would replace libraries, eliminate paper and reduce the rapidly mounting costs of libraries and their growing mountains of paper. The mainframe computer itself served as a kind of ominous, controlling, masculine metaphor for this period. The resulting shift of computing and information systems to new hardware and software platforms, and the growing discovery that centralized, mainframe computing were replacing libraries as the black holes of campus budgets, set the stage for new roles for university library directors.

IV. 1985–present: the "Networked Librarian": or "Change-Agent".

The Age of the Small Machine and the World Wide Web has been good to library directors who are comfortable with innovation, and who understand and experiment with the wealth of new hardware and software toys and tools which become available daily.[10] This most recent period has seen the final demise of the university librarian as defender of the status quo and the rise of the concept of the university librarian as an advocate of progressive and continual improvement in knowledge management. Those who move too far too fast, without analysis of outcome, cost and benefit, have gotten burned, but even for those content with life on the trailing edge, and for those who continually push that trailing edge forward as new tools are tested and proven, this is proving to be an exciting time. It is here that the characterization of the university librarian as "agent-provocateur" becomes appropriate, and my intended meaning explained. By "agent-provocateur" we mean nothing hostile or anti-social, but a role for the library director as teacher, philosopher of values, instigator, innovator and provocative administrator: the librarian as a "change agent" on the university stage.

It is clear that research library directors have learned from their mistakes, and that they have begun to use the tools and concepts of applied social research as well as open systems analysis and statistical representation. They have also become more adept at the use of user and staff-based strategic planning. They have learned that libraries are human organizations, not just factories, and that understanding of the dynamics of organizational communication and learning are as important as management systems and spreadsheets. In addition to learning from the past, and better education for librarians, the understanding of the human and organizational development in university libraries has been speeded and facilitated by the growing number of women who are university librarians and senior managers.

Another key to successful change has been the growing study and understanding of the architecture and culture of digital information, and its web and internet environment. Somehow, for many who we may call "Digital Librarians", the age of the Web, the Internet, and Distributed Information Systems seems familiar, if not entirely comfortable, and they find their way more surely in the distributed, client-server world of the networked pc than they could during the reign of the mainframe. And the plurality, ubiquity and distribution of the means, as well as the ends, of the server-small computer environment, contribute to this process.

This may be the case partially because the digital world of the internet and the web can itself be understood as a metaphor of human society, with its qualities of culture, learning, community and communication. And librarians are librarians – at least in part – because they understand these things and feel a deep commitment to them. Librarians also understand the difference between content which contributes to the outcomes of scholarly communication, and the means or media of communication. At any rate, librarians are good at information management and access, and are constructing an incredibly rich, varied and original, well-organized and accessible new "virtual" library of libraries. The Web Gateway at Penn registered 25 million web pages served last year, growing at an annual rate of nearly 40%, and these figures could be replicated many times over in the research library community.

The changes apparent during this period of the "Digital Librarian" may be characterized by the following list of changes, freely adapted from a recent article by Irene Hoadley[11] (and I acknowledge that some of these were championed or under way before this period began – any periodization is faulty, as well as useful):

• An internal to external role for the director [AND the library]
• Manual to digital, distributed operations

- Hierarchical to flattened organizations
- Internal to distributed access, resources and assistance
- Directives to communication
- One to multiple-format collections
- Simple to complex
- Institutional funding to "multifaceted" funding
- Ownership to access
- Technical skill to four skills: Technical, Human-organizational (team and network-building, negotiational), Conceptual and Innovative-entrepreneuria
- Internal control to user- and outcomes-based services
- Control to delegation and greater participation in organizational decision making
- Organization of structure to leadership of human organizations
- Local to consortial.

The medium may be the message (or the massage), as Marshall McLuhan taught us, but the content is the outcome; and the assessment of outcomes and benefits remains a key goal of current library leaders, along with the standards and data both statistical and anecdotal that will allow benchmarking, comparison and assessment of outputs and use of library information resources and services. Outcomes (and the emphasis on library outcomes may herald yet a new age aborning) also imply careful, inclusive strategic planning that results in values, missions and goals that are broadly understood both within our library organizations and the academic communities they support. And we may note with pleasure the growing exchange and discussion of strategic planning documents and collective strategic planning exercises by teams or consortia of libraries.[12]

NOTES

1. The story is told by Kenneth E. Carpenter, *The First Three hundred Years of the Harvard University Library. Description of an Exhibition.* Cambridge, 1986, p. 68.
2. Rochester, 1876.
3. Epitomized by the famous painting by Charles Willson Peale, in which the artist, as a kind of magus, holds open the curtain to his "cabinet of curiosities", a small natural history and medical museum. The new University of Pennsylvania library opened on its West Philadelphia campus in 1891 as a library-cum-museum of anthropology and archaeology.
4. Guy R. Lyle, *The Librarian speaking; Interviews With University Librarians.* Athens, Ga, 1970, p.vii.
5. The military unit's hierarchical "table of organization and equipment" epitomized the model for many library organizations.

6. Richard M. Dougherty and Fred J. Heinritz, *Scientific Management of Library Operations*. N. Y. and London, 1966, p. 19.

7. Bombay, 1935.

8. Richard M. Dougherty, "Achieving Preferred Library Futures in the 1990's: What is required?, *Research Libraries: Yesterday, Today, and Tomorrow*, W. J. Welsh (Ed.). Westport, 1993, p. 49.

9. *Library Journal*, *100*(1975), 1476.

10. Innovation is used here to mean the generation, acceptance and implementation of new ideas, processes, products and services. V. A. Thompson, "Bureaucracy and Innovation". *Administrative Science Quarterly*, *10*(1965), 1–20.

11. Irene B. Hoadley, "Reflections: Management Morphology How We Got To Be Who We Are". *Journal of Academic Librarianship 25*(1999), 267.

12. I would like to acknowledge Tom Shaughnessy's significant article, "The Library Director as Change Agent". In: J. J. Branin (Ed.), *Managing Change in Academic Libraries 22*(1996) pp. 43–56, for ideas, and for having to devise a different title for this paper.

ABOUT THE CONTRIBUTORS

Theresa S. Byrd is Director of Libraries at Ohio Wesleyan University. She served as Director of Learning Resources at J.Sargeant Reynolds Community College (JSRCC) from 1986–1998. In 1994, She received the Virginia Library Association's George Mason Award for outstanding work for libraries within the Virginia Community College System. She received her M. L. S. from North Carolina Central University, her M.Ed. for Virginia Commonwealth University, and her Ed.D. from the University of Virginia.

Paul N. Courant serves as the Associate Provost for Academic and Budgetary Affairs, Arthur F. Thurnau Professor, Professor of Economics and Public Policy, and Faculty Associate in the Institute for Social Research at the University of Michigan. He has served as Chair for the Department of Economics and Director of the Institute of Public Policy Studies. He has also served as a Senior Staff Economist at the Council of Economic Advisers. Courant has authored half a dozen books, and over sixty monographs and papers covering a broad range of topics in economics and public policy, including tax policy, local economic development, gender differences in pay, housing, radon and public health, and relationships between economic growth and environmental policy.

Billy E. Frye became the fourth Chancellor of Emory University in June 1997 following his service as Vice President for Academic Affairs and Provost since 1988. Frye graduated from Piedmont College in 1953 and earned both his MS and Ph.D. degrees from Emory in 1954 and 1956, respectively. He has served as associate editor of the Journal of Experimental Zoology and of General Comparative Endocrinology. Throughout his career Frye has been a long time supporter and members of the board of the Council on Library and Information Resources and a leading advocate of libraries and constructive digital information resources in colleges and universities.

Murle E. Kenerson received his M. L. S. from the University of Michigan in Ann Arbor, and currently serves as the Assistant Director of Libraries and Media Centers at Tennessee State University in Nashville, Tennessee. He has had previous appointments as a Geier Fellow at the Tennessee Board of Regents and the Library of Congress in Washington, D. C. He also worked at the Chicago Public Library and St. Joseph Mercy Hospital, Riecker Memorial Library in Ann Arbor, Michigan.

Paul Mosher is Vice Provost and Director of Libraries at the University of Pennsylvania, where he is also an adjunct professor of History. Before coming to Penn, he was Deputy Director of Libraries at the Stanford University Libraries. Dr. Mosher has also been Chair of the Research Libraries Group Board, had held major office in the American Library Association, and serves on the Board of the Association of Research Libraries. He has published numerous articles in the field of librarianship.

Steve Reynolds is Vice Chancellor for Information Technology at Indiana University East. He came to IU after having served as Chief Information Officer at Nichols College and Director of Information Systems at LaGrange College. He received his Master of Science in Organizational Behavior from National Technological University, and his Ed.D. in Computing and Information Technology from Nova Southeastern University.

Alexander (Sandy) Slade is Head of Document Supply Services at the University of Victoria in British Columbia, Canada. In this position, he manages the provision of library services to off-campus students, other libraries, and non-university clients. Over the past 20 years Sandy has been very active in the area of library services for distance learning programs. He is co-author of the three annotated bibliographies on library services for distance learning, the most recent one published by Libraries Unlimited in 2000.

Elizabeth A. Titus is Dean of the Library at New Mexico State University in Las Cruces, NM. She has a Ph.D. in Political Science and Master's Degrees in Urban Planning and Library Science. Her fields of specialization include library statistics, library facilities, and library administration & management.

Connie Van Fleet is an Associate Professor in the School of Library and Information Studies at the University of Oklahoma. Dr. Van Fleet is the author of numerous publications and is co-editor of Reference & User Services Quarterly. She is active in the Association for Library and Information Science Education and the American Library Association and currently serves on ALA Council. She was the 1996 recipient of the Reference and User Services Association Margaret E. Monroe Adult Services Award.

Danny P. Wallace is Director and Professor in the School of Library and Information Studies at the University of Oklahoma. Dr. Wallace received the 2000 ALISE (Association for Library and Information Science Education) Award for Teaching Excellence in the Field of Library and Information Science Education. He has served on the ALISE Board of Directors and the Board of Directors of Beta Phi Mu, is a member of the Editorial Board for the Journal of the American Society for Information Science and Technology, and is co-editor of Reference & User Services Quarterly.

Robert Wedgeworth currently serves temporarily as Vice Chairman and Interim President of Laubach Literacy International, a literacy training and advocacy group based in Syracuse, New York. He retired as University Librarian and Professor from the University of Illinois at Urbana-Champaign in August 1999. Prior to that, he served as Dean, School of Library Service, Columbia University (1985–1992) and Executive Director of the American Library Association (1972–1985).

KEYWORD INDEX